T0135529

Reconfigurable Architectures and Design Automation Tools for Application-Level Network Security

Sascha Mühlbach

Bibliografische Information der Deutschen Nationalbibliothek

Die Deutsche Nationalbibliothek verzeichnet diese Publikation in der
Deutschen Nationalbibliografie; detaillierte bibliografische Daten sind
im Internet über http://dnb.d-nb.de abrufbar.

Zugl.: Darmstadt, Techn. Univ., Diss., 2014
D17

ISBN 978-3-8325-3955-9

Logos Verlag Berlin GmbH
Comeniushof, Gubener Str. 47,
10243 Berlin
Tel.: +49 (0)30 42 85 10 90
Fax: +49 (0)30 42 85 10 92
INTERNET: http://www.logos-verlag.de

Abstract

The relevance of the Internet to all parts of business and private use has dramatically grown in the past decades. Social networks etc. change the way how people communicate with each other. Online banking, online shopping and eGovernment solutions simplify day-to-day tasks. Furthermore, even traffic control systems, energy transport networks and other critical infrastructure systems are connected today using the Internet, instead of proprietary closed networks. And this trend will continue.

However, on the flip side, the enormous financial relevance attracts many types of criminals. The worldwide access to systems allows them to remotely commit criminal activities without being on site. Attackers exploit weaknesses (bugs, design errors etc.) of computer systems to break into someone else's systems and, e.g., takeover the machine for other attacks, steal sensitive data or damage physical infrastructure (e.g., power plants) that is controlled by the system. As shown by numerous studies, the risk is omnipresent.

Setting up proper security mechanisms (e.g., Firewalls and Intrusion Detection Systems (IDS)) has therefore never been more important than today. However, the data volume transferred on today's high-speed networks (10 Gbit/s is standard for a backbone network) presents a significant challenge to current security measures. Conventional software-programmable processors are not able to keep up with these speeds. To support such network security mechanisms on the infrastructure level, dedicated hardware accelerators have been proposed to offload compute intensive tasks from the processor.

As a key technology for this approach, Field Programmable Gate Arrays (FPGAs) are of particular interest to this end. FPGAs are integrated circuits whose functionality is not fixed during the manufacturing process, but instead can be flexibly reconfigured for specific applications afterwards. Hardware accelerators based on FPGAs have often been employed for computation or lower-layer acceleration tasks on the packet level. However, advances in chip design and fabrication technology have led to very powerful reconfigurable devices that have become capable of performing more complex, higher-level application operations.

Such higher-level applications will be addressed in the current thesis. In this context, the term high-level defines any system that actively takes part of a network communication session as an endpoint instead of simply monitoring traffic flying by. To support rapid prototyping of such interactive communication applications directly on dedicated hardware, a novel FPGA platform (called NetStage) has been designed and developed. NetStage provides a communication core for Internet communication and a flexible connection bus for attaching custom modules (called Handler) that supports internal routing in a way of network sockets. Furthermore, NetStage contains service modules, e.g., to store application state information at a central place, and supports dynamic partial reconfiguration of Handlers. To ease the development of applications, a simulation environment is provided that can be directly connected to live network traffic.

Altogether, NetStage delivers a rapid prototyping environment where designers can focus on building the application logic and reuse existing infrastructure components. But in addition to the performance aspects, dedicated hardware has another interesting aspect. The complete removal of all software-programmable processors protects the system from subversion when an attacker attempts to inject malicious code. Therefore, NetStage implements all its functionality (from packet processing up to the application layer) as dedicated hardware blocks.

As a proof-of-concept application, the hardware honeypot MalCoBox has been developed for the NetStage platform. A honeypot is a network security device that emulates thousands of vulnerable servers and is placed at an exposed position in the network in order to attract attackers and is therefore a good example for a high-throughput but security-critical application. Honeypots are, e.g., used to deflect attackers from real production services or to gain knowledge about attack patterns.

Despite the advantages of hardware-accelerated platforms, such as NetStage, the high programming complexity of dedicated hardware systems is an ongoing issue. This is especially true for applications such as the MalCoBox, where the developers will be domain experts in network security, but not proficient in the digital logic design and computer architecture fields commonly required to program FPGA-based computers. To resolve this quandary, the domain-specific Malacoda language for abstractly formulating honeypot behavior is presented and discussed.

An associated compiler translates Malacoda from a high-level description of the honeypot's packet communication dialogs into high-performance hardware handlers. In order to reach the required performance, the compiler follows a different paradigm compared to conventional approaches: the operations in the program are not mapped to byte-wise copy-and-select steps when executing the program. Instead, they are turned into a dedicated wiring/logic network that modifies all bytes of a data word in parallel. In this manner, the generated handlers are able to keep up with the stream of network packets in real time. Together, NetStage and Malacoda address some of the productivity deficiencies often recognized as major hindrances for the more widespread use of reconfigurable computing in communications applications.

Finally, the NetStage platform has been evaluated in a real production environment. An instance of the MalCoBox running on NetStage installed on a NetFPGA 10G FPGA development board has been connected to a 10G University uplink for one month, collecting various statistics about core network interoperability and the honeypot interactions, demonstrating the practical usability of the described architecture.

Kurzfassung

Die Bedeutung des Internets für alle Bereiche des privaten und geschäftlichen Lebens hat sich in den letzten Jahren extrem gesteigert. Soziale Netzwerke und ähnliche Communities verändern die Art und Weise, wie Menschen miteinander kommunizieren. Online Banking, Online Shopping und eGovernment-Lösungen vereinfachen die Abläufe unseres täglichen Lebens enorm. Zudem werden auch immer mehr kritische Infrastruktursysteme (z.B. für Verkehrssteuerung oder die Energieverteilung) mit dem Internet verbunden, anstatt sie über proprietäre abgeschottete Netze zu verwalten. Und dieser Trend wird sich weiter fortsetzen.

Die Kehrseite dieser Entwicklung ist, dass die enorme finanzielle Relevanz eine ganze Reihe von kriminellen Aktivitäten anzieht. Durch die weltweite Zugriffsmöglichkeit auf Systeme müssen sich die Verursacher zudem noch nicht einmal mehr in der Nähe befinden. Angreifer nutzen Schwachstellen in Computersystemen (z.B. Bugs), um in entfernte Rechner einzudringen und dieses Systeme entweder für Angriffe auf weitere Systeme zu verwenden, sensible Daten zu kopieren oder mit den Systemen verbundene physikalische Infrastrukturen zu zerstören (z.B. Produktionsanlagen). Dieses Risiko ist allgegenwärtig, wie immer wieder durch entsprechende Meldungen und Studien belegt wird.

Der Einsatz von passenden Sicherheitsmechanismen (z.B. Firewalls oder Intrusion Detection Systems (IDS) ist unausweichlich. Allerding stellen die hohen Datenraten in modernen Kommunikationsnetzen (10 Gbit/s sind mittlerweile Standard für Netze im Backbone) eine große Herausforderung für solche Mechanismen dar. Konventionelle software-programmierbare Prozessoren können hier mit diesen Datenraten nicht mithalten. Für den Einsatz auf der Infrastrukturebene wurden daher dedizierte Hardwarebeschleuniger eingeführt, welche den Prozessor von rechenintensiven Operationen entlasten.

Eine wichtige Rolle spielen hier die so genannten Field Programmable Gate Arrays (FPGAs). FPGAs sind dedizierte Chips, deren Funktionalität nicht bereits im Fertigungsprozess final festgelegt wird, sondern die sich später flexibel verändern lassen. Hardwarebeschleuniger auf Basis von FPGAs wurden in der Vergangenheit häufig für die untere Netzwerkschicht auf Paketebene eingesetzt. Allerdings hat die Entwicklung im Bereich dieser FPGA-Technologie mittlerweile so leistungsfähige Bausteine hervorgebracht, die sich auch für Operationen auf höheren Kommunikationsebenen eignen.

Der Einsatz von FPGAs für genau solche höherschichtigen Anwendungen wird in der vorliegenden Arbeit behandelt. Der Begriff "höherschichtig" umfasst in diesem Kontext alle Systeme, die aktiv an einer Kommunikationssitzung als

Endpunkt teilnehmen, anstatt nur passiv vorbeilaufende Datenpakete zu untersuchen. Um die schnelle Prototypentwicklung solcher interaktiven Systeme auf dedizierter Hardware zu ermöglichen, wurde eine neue FPGA Plattform (genannt NetStage) entwickelt. NetStage stellt einen zentralen Kommunikationskern für die Kommunikation im Internet zur Verfügung und ermöglicht die Anbindung applikationsspezifischer Module (genannt Handler) über einen flexiblen Bus, der ein internes Routing vergleichbar mit Netzwerksockets abbildet. Zudem beinhaltet NetStage weitere Hilfsmodule, wie z.B. einen zentralen Sessionspeicher für Applikationen, und unterstützt die dynamische partielle Rekonfiguration von FPGAs für einzelne Handler. Zur weiteren Vereinfachung der Arbeit mit der Plattform wird eine Simulationsumgebung bereitgestellt, die sich interaktiv an reale Netzwerkdaten ankoppeln lässt.

Zusammenfassend bietet NetStage eine schnelle Entwicklungsumgebung, die fertige Infrastrukturkomponenten als Grundlage bietet und sich der Entwickler somit auf das Design der eigentlichen Logik einer Applikation konzentrieren kann. Zusätzlich zu den genannten Geschwindigkeitsvorteilen bietet dedizierte Hardware noch einen weiteren interessanten Aspekt. Wenn man jeglichen software-programmierbaren Universalprozessor aus dem Daten- und dem Kontrollpfad entfernt, so kann ein solches System von einem Angreifer nicht mehr verwendet werden, um schadhaften Code auszuführen. NetStage implementiert daher jegliche Funktionalität (von der Paketverarbeitung bis zur Applikationslogik) als dediziertes Hardware-Modul.

Ergänzend zur Entwicklung der Plattform, wurde ein hardware-basierter Honeypot als Machbarkeitsstudie für die NetStage Plattform implementiert (genannt MalCoBox). Ein Honeypot gehört zur Klasse der Netzwerksicherheitskomponenten und emuliert tausende verwundbare Applikationen, die an exponierter Stelle im Netzwerk platziert werden, um potentielle Angreifer anzulocken. Dadurch bietet er ein gutes Beispiel für eine durchsatzstarke, aber zudem aufgrund ihrer exponierten Platzierung selbst höchst gefährdete Applikation im Netzwerk. Honeypots werden aktiv verwendet, um Angreifer von den echten Produktivsystemen abzulenken und um die Vorgehensweisen der Angreifer zu studieren.

Obgleich die Verwendung von hardwarebeschleunigten Plattformen (wie NetStage) erhebliche Vorteile mit sich bringt, ist die komplexe Programmierung solcher Systeme ein anhaltendes Problem. Dies zeigt sich im Besonderen bei Applikationen wie der MalCoBox, wo die Entwickler in der Regel Experten auf dem Gebiet der Netzwerksicherheit sein werden, nicht jedoch zwingend Erfahrung im Design digitaler Schaltkreise und Computerarchitekturen mitbringen, was jedoch für die direkte FPGA-Entwicklung unerlässlich ist. Um dieses Dilemma zu umgehen, präsentiert und diskutiert diese Arbeit die anwendungsspezifische Sprache Malacoda, die es erlaubt das Verhalten des Honeypots auf abstrakter Ebene zu formulieren.

Ein dazu entwickelter Compiler übersetzt die Malacoda-Beschreibungen in hochperformante Hardware-Module für die NetStage Plattform, die die Kommunikationsdialoge des Honeypots ausführen. Um die geforderte Geschwindigkeit zu erreichen, setzt der Compiler auf ein von konventionellen Vorgehensweisen abweichendes Paradigma: Die Operationen des Malacoda-Programms werden nicht zu einzelnen Schritten übersetzt, sondern der Compiler orientiert sich an den parallelen Anforderungen für jeweils ein Datenwort und erzeugt hierfür einen dedizierten Schaltkreis. Hierdurch kann ein so erzeugter Handler in Echtzeit mit dem Paket-Datenstrom arbeiten. Insgesamt werden von NetStage und Malacoda eine Reihe von Defiziten adressiert, die trotz der Vorteile immer wieder als Hindernis für eine weitere Verbreitung von rekonfigurierbaren Hardware-Technologien wie FPGAs im Netzwerkkommunikationsbereich genannt werden.

Für eine abschließende Bewertung wurde die NetStage Plattform in einem realen Produktivumfeld getestet. Eine Instanz der MalCoBox basierend auf der NetStage Plattform wurde auf einer NetFPGA 10G FPGA Entwicklungskarte installiert und für einen Monat an einen 10G Uplink eines universitären Rechenzentrums angeschlossen, um verschiedene Ergebnisse über die Interoperabilität der Kernplattform und die Interaktionen des Honeypots zu sammeln. Hierdurch wurde die praktische Verwendbarkeit der vorgestellten Architektur demonstriert.

Acknowledgments

First of all, I would like to thank Prof. Andreas Koch for his excellent and helpful support over the past years by always being available for discussions about ideas, architecture, and implementation details for the various aspects of my dissertation. Beyond that, he provided me with the required equipment and together we were able to set up valuable external cooperations. Overall, these aspects were important factors for the success of my work.

Next, I would like to thank Prof. Mladen Berekovic for accepting to be the second referee of this work.

Furthermore, I would like to thank Wolfgang Hommel and Stefan Metzger from the Leibniz-Rechenzentrum München for giving me the opportunity to test my hardware honeypot application in a real datacenter environment. Additionally, I would like to thank Peter Schoo and his team from the Fraunhofer AISEC München for helpful discussions about honeypots and my new domain specific language. I am also thankful to Michaela Blott and Gordon Brebner from Xilinx Research for their assistance, particularly with the NetFPGA 10G prototype board.

In addition, I am grateful to my colleagues from the Center for Advanced Security Research Darmstadt and the Embedded Systems and Applications Group at the TU Darmstadt for the good cooperation and a nice time together.

Finally, my special thanks go to my wife Heike, always supporting my project.

This work was partly funded by LOEWE CASED.

Contents

1 Introduction

The relevance of the Internet, the worldwide network of computers, to all parts of business and private use has dramatically boosted in the past decades. E.g., the sheer number of Internet hosts counted by the Internet Systems Consortium (ISC) ramped up from 213 in 1981 to roughly 888,000,000 in 2012 [ISC12].

Nowadays, the Internet has a huge impact on all areas of our personal live. In 2009, an estimated 1.7 billion citizens used the Internet (growing 380 percent from 2000 to 2009) [Atk10]. Social networks, private clouds etc. change the way how people communicate with each other. Online banking, online shopping and eGovernment solutions simplify day-to-day tasks. Furthermore, even traffic control systems, energy transport networks and other critical infrastructure systems are today connected using the Internet instead of proprietary closed networks. In the future, this will continue. E.g., connecting smart devices, such as sensors in houses, cars etc. (also known as the Internet of Things [Atz10]), will be one of the next major trends.

Apart from private activities, a huge Industry sector has grown in that domain. So called Internet companies base their entire business model on the Internet (e.g., Google, Facebook). However, the enormous economic benefit from near-realtime access to systems and business partners all over the world does also affect the traditional industry sectors (e.g., manufacturing) by allowing them to efficiently communicate with suppliers or outsourcing partners. The ITIF estimates that the annual global economic benefits of the commercial Internet equals $1.5 trillion [Atk10].

However, on the flip side, the enormous financial relevance attracts many types of criminals. The worldwide access to systems allows them to perform criminal activities without being on site. Attackers exploit weaknesses (bugs, design errors etc.) of computer systems to break into someone else's systems and, e.g., takeover the machine for other attacks, steal sensitive data such as passwords, credit card information, engineering drawings etc., or damage physical infrastructure (e.g., power plants [CER12]) that is controlled by the system. The risk is omnipresent, as studies from large Internet security companies show every year [Sym12, Sop12].

Malware, such as viruses or worms, that automatically exploits bugs of computer systems to execute malicious code written by the attacker, is a common tool for starting attacks. Pandalabs estimates that in 2012 32 percent of the worldwide PCs are infected with some malware [Pan12b]. In some cases, such malware infects the victim and silently copies data for months until it receives attack commands from the Internet, so that it can stay there undiscovered for a long time. Such so called advanced persistent threats (APT) [Zet10] are hard to defeat and can result in massive losses. With their increased relevance also embedded devices, such as payment terminals [Onl12] or smartphones [Dav12], can be targets of remote attacks.

Setting up proper security mechanisms has therefore never been more important than today. E.g., a virus scanner is an efficient tool for PCs to defeat the majority of malware attacks, which is strongly recommended for any user. Firewalls and Intrusion Detection Systems (IDS) are two technologies that are employed on the network layer to secure the communication infrastructure from illegal access to systems and applications, so that attacks cannot reach the target systems. However, the data volume transferred on today's high-speed networks (10 Gbit/s is nowadays standard for a backbone network and 40 Gbit/s and 100 Gbit/s links are under development) presents a significant challenge to current security measures. Conventional software-programmable processors are not able to keep up with these speeds. An evaluation of the popular network IDS Snort showed that such a software system can scan only up to 200 Mb/s without noticeable packet loss on a regular server system [Als09].

To support such network security mechanisms on the infrastructure level, dedicated hardware accelerators have been proposed to take away compute intensive tasks from the processor. While so-called network processors [Lee04a], that contain programmable processing elements dedicated for packet processing and offer a reasonable tradeoff between performance, programming complexity and costs, are dominant in the commercial sector, the research community also focuses on reconfigurable hardware, namely Field Programmable Gate Arrays (FPGAs), that allows to develop custom dedicated hardware to achieve outstanding performance for particular scenarios. FPGAs are integrated circuits, whose functionality is not fixed during the manufacturing process, but instead can be programmed afterwards. This makes them very attractive for low volume applications, prototyping and research. However, programming them requires in-depth knowledge of digital system design.

While such hardware accelerators based on FPGAs have been mainly employed for lower-layer acceleration tasks on the packet level during the past years [Kat07, Jed08], the development in chip technology brought up very powerful devices, that are now ready to take over more complex higher-level functionality. Such higher-level applications will be addressed in the current thesis. In the following, the term high-level defines any system, that actively takes part of a communication session as an endpoint instead of simply monitoring traffic flying by (as most of the IDSs do). To support rapid prototyping of such interactive communication applications directly on dedicated hardware, a novel FPGA platform (named NetStage) has been designed and developed as part of this work. This platform allows testing new ideas quickly and taking measurements under real conditions. The relevance of this topic is also underlined by a recent article of a senior research engineer from Xilinx [Blo12], one of the two global leading FPGA vendors, that states that FPGAs could take an important role for the future cloud applications if proper design environments can be provided.

The implementation of network-security related applications on dedicated hardware benefits from two major advantages of hardware-based systems: performance and resilience. As said, dedicated processing elements allow the development of high-performance compute engines, which outperform general purpose processors by multiple orders of magnitude. But additionally, dedicated hardware is more resilient against compromising efforts. As there is no processor that can be abused to run malicious code, there is no point of attack for malware.

For an evaluation of the platform, a prototype of a hardware-only honeypot has been developed as part of this thesis. A honeypot is a network security device that emulates thousands of vulnerable servers, placed at an exposed position in the network to attract attackers attacking them. The goals are to deflect attackers from real production services and to gain knowledge about attack patterns. The high volume of requests that could reach such a honeypot when connected to a high-traffic link together with the increased risk of such an exposed system to get attacked and abused itself, turns it into an interesting candidate for a hardware implementation.

In addition to the raw hardware platform, this work covers another important aspect of using FPGAs in the field: programming complexity. As one could not take knowledge in digital system design as granted for every network researcher, there needs to be another way to make such a hardware honeypot usable for them in a day-to-day fashion (e.g., frequent updates of emulations). At the example of the honeypot, a high-level language, named Malacoda, for describing the honeypot behavior in an abstract way has been designed. A corresponding developed compiler automatically transforms the high-level description into high-performance hardware for the NetStage platform. Finally, the honeypot prototype system (named MalCoBox [Müh12b]) has been connected to a real Internet uplink to gather real-world results.

The remainder of this chapter will now introduce the three previously sketched major areas of work that cover relevant parts of this thesis.

1.1 Hardware Support for Network Security

In practice, security mechanisms for networked devices can be applied on multiple levels. They can protect the software running on the device (e.g., application monitors), the operating system (e.g., virus scanners), the network (e.g., firewalls or IDSs) or the hardware itself (e.g., protection against side channel attacks). Efficient security strategies for an environment need to combine elements from all available levels. FPGAs are well suited to support mechanisms on the network layer (see below). This is also known as network infrastructure security and covers devices such as firewalls, proxies, and intrusion detection and prevention systems.

Dedicated hardware (FPGAs as well as ASICs) achieves its speed gain over the traditional von-Neumann architecture (that is the foundation of most software-programmable CPUs), by massively exploiting instruction level parallelism and pipelining, as well as the absence of instruction fetch cycles [Sir08]. This is especially beneficial for compute intensive applications or algorithms that can be well parallelized. A good example is the acceleration of the network IDS Snort [Roe99], that is the de-facto standard in the open source community.

An IDS inspects packets for suspicious content defined by regular expressions that match particular string or byte sequences. The core of such an application is the implementation of fast regular expression matching units, something which can be easily parallelized on hardware (e.g., [Sou04, Lo09, Wan10]). Note that IDSs detect intrusions, but do not perform any further actions on the packets. But such a system can be combined with a dynamic packet filter to form an intrusion prevention system [Wea07].

In addition to the pattern- and rule-based IDSs, nowadays also anomaly-based IDSs are gaining higher attention. While the former do primarily detect attacks where a signature is previously known, the latter apply special algorithms that detect deviations in network behavior from a baseline and alarm suspicious activities, without actually knowing in detail what type of attack it is. Such anomalous behavior can be, e.g., an unexpected peak in outgoing traffic, which can mean that someone is stealing data. A crucial part for anomaly detection is the collection and aggregation of network statistics at high network speeds.

To get accurate results, no packet should be left uncounted. Here, FPGAs can be employed to easily inspect and classify packet streams at 10 Gbit/s or more [Jia09].

Research on accelerating IDSs has a significant role in academia. However, previous work does focus on packet level inspection and only rarely addresses the entire implementation of such applications on dedicated hardware. This is especially true for the high-level communication endpoints. This domain has been traditionally dominated by software-programmable devices, obviously because of their easier programmability. Furthermore, implementing active communication on hardware requires a hardware-only version of the basic Internet communication protocols (IP, TCP etc.), which has not been addressed in the research community for 10 Gbit/s or more in the past years. This gap will be closed with the current work by the implementation of a 10G Internet communication stack for FPGAs that supports easy prototyping of hardware-only Internet communication applications.

One example of a network security application that has been traditionally implemented on software systems is a honeypot. Low-and medium interaction honeypots emulate basic operations of an application, e.g., to distract an attacker from the actual production systems (and generate an early warning that a potential attack has been detected on the honeypot) or to gather malware samples automatically, while high-interaction honeypots provide entire operating systems with applications to attackers (e.g., to study the steps an attacker takes to compromise the system to muster a proper defense on other systems). With their well-defined set of operations, especially low- and medium interaction honeypots are ideal candidates for a prototype application to be transferred to hardware. The implemented honeypot (the MalCoBox) is one of the first published hardware-only network infrastructure security devices in that domain.

1.2 Platforms for Hardware-Based Networking

In addition to a core communication stack, prototyping entire applications additionally requires some type of infrastructure to stick together the single components and control the execution of operations. Furthermore, due to the long development cycles for FPGA designs (see Section 2.1.2), efficient hardware development requires special simulation and debug capabilities. Providing a set of assistant tools along with the infrastructure is therefore of additional importance. Lastly, the availability of suitable (and affordable) development boards equipped with FPGAs is an important point.

Fortunately, the relevance of hardware-based networking, especially with FP-GAs, motivated a couple of research groups and commercial vendors to provide suitable hardware boards for easy system development. One of the most famous platforms in this area is the Stanford NetFPGA [Loc07]. The NetFPGA base hardware (see also Section 2.2.2) is available in either a 1G or a 10G [Net11b] version and has an FPGA, four network interfaces, and various external components (e.g., RAM and Flash memory). The hardware is sponsored by major semiconductor companies, thus achieving a very attractive price for educational use, so that NetFPGA cards are used for a broad range of research activities.

Apart from the raw hardware, NetFPGA is also an open source project that provides a huge set of modules, sample applications, development tools and documentation for free. This allows researchers and especially students to quickly start working with FPGAs and networking. Due to this, an active community has spin around NetFPGA, providing various third-party projects [Net12b], mostly in the domain of switching and routing.

While from a first look NetFPGA might be a good candidate as foundation for this work, it's initial focus on the switching and routing applications [Loc07] (that inherit a flow-based communication that is different from the endpoint-oriented communication required for the applications targeted here) and the lack of a communication stack hinder to take the NetFPGA codebase directly. Furthermore, the 10G codebase has just recently been made available as public beta [Net12a] and was not available when starting with this work.

As there was also nothing suitable on the commercial sector (the COMBO boards from INVEA-TECH [IT10] looked promising, but did also not contain a communication stack as part of their software suite NetCope [Mar08]), a new software platform has been developed from scratch: NetStage [Müh10a]. NetStage provides a communication core (for Internet communication) and a flexible connection bus for attaching custom modules that supports internal routing in a way of network sockets.

Furthermore, NetStage contains service modules, e.g., a Global Application State Memory, and supports Dynamic Partial Reconfiguration (DPR), a technology that allows to change the behavior of an FPGA during operation. Furthermore, a simulation environment eases the development of applications. Altogether, NetStage delivers a rapid prototyping environment where designers can focus on building the application logic and reuse existing infrastructure components. To benefit from the discounted NetFPGA hardware, the NetFPGA 10G hardware board is supported as target hardware for NetStage.

1.3 High-Level Compilation of Networking Hardware

A hurdle for hardware based systems is that programming them still requires special expertise in digital system design. When developing FPGA applications, in addition to the application logic, clock synchronization and timing issues are major concerns that are not familiar to software developers. Furthermore, the high amount of parallel activity on a hardware chip requires good knowledge of parallel programming concepts, which in fact has not been in the focus during the past years due to the long-lasting dominant single-core architecture of computers. However, there have been some changes in this direction due to the now available multi-core processors.

All this results in hardware development being still more time-intensive and costly as developing software. This is especially detrimental when building applications such as the honeypot, where frequent updates (e.g., new vulnerability emulations) are required. Therefore, some type of high-level abstraction of FPGA programming is required here. In general, this issue is not limited to the domain of hardware-based network security, but instead a large blocker for the broad use of FPGAs in commercial and academic applications. Because of this, high-level compilation of hardware is one of the major topics in research and industry.

One proposed approach is to use a general programming language (GPL), e.g., C, as the base language and compile hardware units out of software programs [Gad07]. However, as FPGAs are not generic processors, there are always some limitations that prevent direct hardware generation out of such generic languages [Edw05]. Therefore, instead of compiling the whole application to an FPGA, heterogeneous approaches combine a software-programmable central processing unit (CPU) and an FPGA on a single system and compilers try to divide the workload between the CPU and the FPGA in a way, that each component can achieve the maximum performance [Koc10]. While this does work very well for particular applications, the use of a CPU should be avoided for security reasons in the current context.

An option is to extend a GPL with special commands for packet processing, as, e.g., packetC [Dun09] proposes. But in such a case, still general programming knowledge is required and efficient compilation requires to program using specific constructs. Instead of using a GPL, another approach is to use a domain-specific language (DSL) that supports an even higher abstraction level and allows formulating problems in the domain of the problem, not on the technical level. Such domain-specific languages can perfectly hide the constraints and limitations of the dedicated target hardware and generate hardware modules that perfectly fit to the application domain. An example is the language G [Bre09] for packet level header processing. Such solutions achieve a much higher productivity and compilation efficiency for particular problems than generic solutions, but their flexibility is of course limited.

In case of the desired honeypot implementation, updates will have a common structure (a module responds to received packets), but the operations itself can be very different (regular expressions, copy operations). Additionally, hardware complexity should be completely hid from the user and high-performant modules for 10G+ operation should be generated. Following this, a DSL (named Malacoda) in combination with a compiler generating specific dedicated hardware has been chosen as best fitting option here. Malacoda allows to describe the packet communication dialog of an application emulation, that can be be compiled into high-speed hardware.

1.4 Thesis Contributions

This work covers research aspects that are important for the design of a development platform that supports research on Internet communication on dedicated hardware providing a secure environment. The main contributions are:

- NetStage: A base platform architecture that supports endpoint-oriented communication on FPGA-based systems using dedicated hardware without any CPU in the main data path for security reasons.

- A lightweight design of an Internet communication protocol stack (IP, UDP, TCP) that is tailored to the needs of the hardware platform.

- A fully working implementation of the platform components actually tested on two hardware boards with a Virtex 5 FPGA and 10G network interfaces (BeeCube BEE3 and NetFPGA 10G).

- Tools to manage the platform and a real-time simulation environment with a virtual network interface for live debugging.

- Support for autonomous DPR based on traffic characteristics to increase the utilization of a single FPGA.

- The MalCoBox: An implementation of a fully working hardware honeypot as showcase and evaluation application for the platform architecture and the single components.

- Malacoda: A domain-specific language for high-level description of honeypot modules for the MalCoBox in an abstract way without requiring knowledge of digital system design.

- A Malacoda compiler that automatically generates high-performance hardware modules to be run on the NetStage platform.

- Results from a live-evaluation of the hardware honeypot connected to a public high-speed Internet uplink in a datacenter for one month.

Some aspects of these contributions have also been published in conference proceedings and as journal articles:

- [Müh10a] presents the initial idea of the FPGA platform *NetStage* for high-level network processing and the honeypot *MalCoBox* as demonstration application. It further introduces the concept of *Handlers* and especially the Vulnerability Emulation Handler (VEH) for the honeypot. The prototype, which was developed on the BEE3 platform, was limited to process only UDP packets.

- [Müh10c] introduces the *lightweight TCP implementation* that has been developed for NetStage. Furthermore, it discusses the protocol variations with respect to standard TCP in greater detail. A basic Web Server emulation handler developed directly in VHDL was used to test the TCP functionality.

- [Müh10b] (invited for further journal publication) discusses the basic implementation of using FPGA *partial reconfiguration* for single Handlers. The corresponding prototype allows to transfer FPGA configuration files for a single Handler via network to the FPGA and then initiate the reconfiguration process locally on the chip.

- [Müh11b] (Best Paper Award, invited for further journal publication) introduces the idea of a *domain specific language* for Handler development (at this time called vulnerability emulation description language, VEDL). The focus lies on the honeypot application example, with more details on the internal Handler operational concept being presented.

- [Müh11a] (invited for further journal publication) presents an implementation capable of *dynamic partial reconfiguration* (DPR) on the NetStage platform to time-multiplex the hardware resources for Handlers. The developed system allows to autonomously exchange Handlers in order to react to changing network traffic characteristics. The FPGA configuration data streams are stored in an external memory on the platform.

- [Müh11d] discusses the use of *multiple FPGAs* on the BEE3 board to extend the Handler capacity of the NetStage platform. The multiple FPGAs are connected using a ring-structured bus, where one FPGA acts as master that manages the connection to the external network, with the remaining FPGAs acting as Handler-processing nodes.

- [Müh12c] extends [Müh11a] by introducing the concept of *passive Handlers* to widen the possible scope for using NetStage. The concept is demonstrated by an example implementation of a basic network monitor Handler. The Handler is able to count network packets destined for particular ports in given time intervals and compares these statistics to a baseline that has been previously stored.

- [Müh12a] extends [Müh10b] by combining the use of partial reconfiguration and multiple FPGAs in a single design. This allows to partially reconfigure Handlers in each of the connected FPGA nodes, while the entire process is controlled by the single master FPGA that has the connection to the external network. Despite the large Handler capacity and high configuration bandwidth (using multiple ICAP ports in parallel), this concept has not been followed up in more recent work due to the high complexity of this design and the very limited availability of large multi-FPGA boards.

- [Müh11c] extends [Müh11b] by examining the use of dynamic partial reconfiguration in NetStage to *virtualize* the FPGA for the hardware honeypot.

- [Müh12b] discusses further developments (based on [Müh11b]) of the domain-specific programming language *Malacoda* for Handler development in greater detail and presents example Handlers written in this language. The NetStage platform has been ported to the NetFPGA 10G board and the results of a live evaluation of the hardware honeypot MalCoBox are described.

- [Müh14] surveys the current state of the entire NetStage / MalCoBox project by summarizing parts of this thesis. It further examines the integration of the architecture into the NetFPGA 10G environment and sets particular focus on crucial details of the Malacoda compiler (e.g., micro-architecture and optimizations).

1.5 Thesis Structure

The structure of this thesis is as follows:

Chapter 1   introduces the domain of hardware acceleration for network security, presents current research problems and the motivation for this thesis, and states the contributions provided by this research.

Chapter 2   introduces the technology of reconfigurable hardware, with special focus on the field programmable gate array (FPGA) that is used as a core element for this work. Furthermore, two FPGA-based hardware boards suited for network research are described, that have been used to develop the prototypes mentioned in this thesis.

Chapter 3   discusses related work in the fields of hardware support for network infrastructure security, development platforms for hardware-based network applications, high-level languages and their compilation into dedicated hardware, and honeypots.

Chapter 4   presents the core architecture of the new NetStage platform and its implementation. This includes especially the lightweight communication stack, message routing and module connectivity.

Chapter 5   describes additional infrastructure that is build around the NetStage core components to arrive at an integrated development platform. This includes the unified module interface, system management, simulation, and platform scaling using DPR. Furthermore, details on application development are given.

Chapter 6   introduces the domain-specific language Malacoda that has been developed to describe the behavior of honeypot service modules. Furthermore, a corresponding hardware compiler is presented, that generates hardware modules for the NetStage platform out of these high-level Malacoda descriptions.

Chapter 7 shows experimental results gathered from the implementation of the
NetStage platform and from a one month lasting live evaluation of the MalCoBox
connected to a real Internet uplink.

Chapter 8 summarizes the content of this thesis and recapitalizes its contributions. It concludes with an outlook towards further research on both the hardware
and the compiler track.

2 Reconfigurable Hardware

In the majority of cases, electronic devices are developed using software running on a general purpose processor (GPP, also commonly named central processing unit (CPU) or simply processor) or by using dedicated hardware for a particular application, so called application specific integrated circuits (ASICs). While software-programmable devices provide nearly unlimited possibilities to implement different applications, ASICs are fixed to a particular solution during manufacturing and cannot be modified later. However, in turn dedicated hardware offers higher performance and lower power consumption due to highly optimized hardware structures. But developing ASIC solutions is a time consuming, expensive and risky process due to the difficult physical process steps that are required to create an integrated circuit. Often such designs easily cost millions of dollars [Hau07] before an ASIC is ready for production and therefore do only make sense for, e.g., mass production or very specific application domains.

To overcome some of these shortcomings, basic programmable logic devices (PLD) have been on the market for some decades now. Such devices contain dedicated logic, but instead of an ASIC, where the logic functionality and interconnectivity is defined during manufacturing, flexible structures allow to be later programmed to implement arbitrary logical functions. However, these basic PLDs were not able to replace ASICs for larger designs. Just the invention of the Field Programmable Gate Array (FPGA) in 1984 by Bernie Vonderschmitt, Ross Freeman, and Jim Barnett [Sta04], the founders of Xilinx, provided a technology that is able to keep up with the complexity of larger designs that were naturally bound to ASICs.

2.1 Field Programmable Gate Array (FPGA)

An FPGA contains a mix of reconfigurable logic blocks and routing resources. As the base logic structures are still implemented as dedicated hardware, FPGAs inherit most of the advantages from ASICs (performance, power efficiency etc.). While being a bit slower than ASICs, FPGAs can still be hundreds of times faster than software implementations [Hau07] due to the parallel computing resources available on the chip. Nowadays, the market for FPGAs is dominated by the two major companies Xilinx and Altera [Xil12e], that together have a market share of over 80%.

2.1.1 Architecture

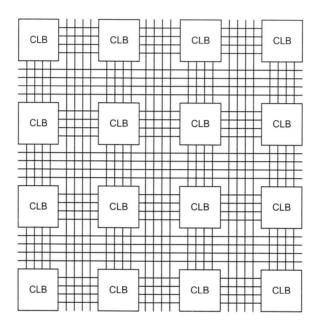

Figure 2.1: Basic FPGA architecture (simplified)

Figure 2.1 shows a schematic overview of the basic FPGA architecture internals. Reconfigurable logic blocks are surrounded with a flexible interconnection structure (also called routing resources). The logic blocks (Xilinx calls them slices) contain resources for implementing combinatorial logic (Boolean equations) as well as sequential logic (Flip-Flops). The Boolean functions inside the logic blocks are often represented using lookup tables (LUTs). Such lookup tables contain the result for any given input values, so that the correct output can be simply looked up.

The routing resources allow to flexibly connect the input and outputs of slices to create a large network of processing elements. LUTs can be easily programmed by changing the content of the lookup tables. Connections between slices are configurable up to a certain extent (e.g., the number of available connections between blocks is limited). The majority of today's FPGAs are based on SRAM technology for storing the control information for slices/LUTs and routing resources. SRAM supports easy programming and the hardware process to create SRAM memory is continuously improved due to the fact that SRAM plays also a major role in other areas of computer design, so that the FPGA vendors can benefit from process improvements here [Max09]. However, SRAM-based FPGAs lose their program when power goes away and need to be reprogrammed on each power cycle.

In contrast, antifuse-based FPGAs can be programmed only once, but they keep their program even when being powered off. In that, their behavior is closer to ASICs. Flash-based FPGAs are somewhere between, as they can be reprogrammed but keep their program when being powered off. The main drawback is that their programming time is multiple times higher than when using SRAMs, which is an issue when using techniques such as DPR (see below). Furthermore, Flash-based devices require more space for the same number of logic resources than their SRAM-based equivalents [Max09].

In addition to simple LUTs and Flip-Flops, modern FPGAs can contain various dedicated resources (digital signal processors, hardware multipliers, on-chip memory, network transceivers etc.), that allow to efficiently use the on-chip resources for complex designs. The Xilinx Zynq-7000 FPGAs [Xil14] can even contain an embedded ARM processor on chip that can be connected with custom application logic to build a reconfigurable System-on-a-Chip (SoC). To connect external devices (memory, processor etc.), FPGAs contain configurable external interface pins.

When comparing FPGA-based designs, basic criteria are the number of logic resources (LUTs and Flip-Flops), as this directly relates to the complexity of the implemented logic, and the achievable operating frequency, which is one of the key factors for performance. Note that due to the fact that some FPGAs can contain various embedded resources (e.g., special RAM or compute accelerators), which might consume regular logic resources on other chips, accurate comparisons between designs made for different FPGA types are somehow difficult and require detailed information of all parameters to make a fair decision.

2.1.2 Development Process

As FPGAs provide programmable dedicated hardware, they are programmed using the same programming languages as for ASICs. The dominant languages are VHDL and Verilog that allow specifying the chip functionality on the register transfer level (RTL). In contrast to procedural programming languages, VHDL and Verilog enforces dataflow-oriented [Kau11] programming, which means that a program describes the data flow between processes. Registers (implemented using Flip-Flops) store intermediate values and transfer data between processes. On FPGAs, all these processes are executed in parallel, thus resulting in the enormous speed gain. However, this parallelism induces issues such as synchronization that need particular attention (which is even more complicated if the processes run at different clock speeds).

An advantage of VHDL and Verilog is that there exist simulation tools that can execute the code and simulate the chip behavior. However, this simulation can be quite difficult, as the state of the chip at a specific point in time is represented by all of its signal values and process states (and not only by some registers and the program counter, as for software running on a GPP). The values need to be recalculated every clock cycle, so that the simulation of one real-time second of a chip running at 100 MHz requires calculating all the values for 100.000.000 clock cycles. Even as there are various optimizations, simulation of hardware is much slower than debugging a program running on a GPP.

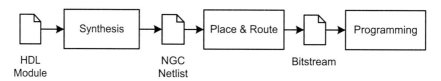

HDL Module   NGC Netlist   Bitstream

Figure 2.2: FPGA design workflow

After a program has been verified in simulation, it can be compiled to an FPGA design for a particular target device. For this, each vendor provides an entire toolchain that controls the various steps involved in that process. The process is sketched in Figure 2.2. First, the RTL description (e.g., a VHDL or Verilog program) is synthesized into a Netlist, that contains a representation of the program mapped to basic FPGA blocks (also called primitives), such as LUTs, Flip-Flops, Memory etc., and the connection information between these blocks. This Netlist is then used as input for the Place & Route (P&R) process, which actually generates the configuration information of the FPGA by distributing the primitives across the available FPGA resources (placement) and establishes connections between the resources using the available routing resources (routing).

However, this process is constrained. As the number of routing resources is limited, logic functions cannot be arbitrarily placed across the chip. Furthermore, e.g., wire lengths of routing lines do have an impact on the operating frequency and in this manner implicitly on the performance. Therefore, the designer can set preferences. E.g., if the goal is a very high speed, FPGA resources are distributed in a different way than if a high packing density is required (which in turn might reduce the overall clock speed). For some applications, designers can also fix the placement of primitives on the FPGA (e.g., required for partial reconfiguration), which further lowers the degree of freedom for the process. As the FPGA routing problem is known to be NP-complete [Wu94], in practice heuristics are used to search for an optimum. But this process can take long times (from hours to days for large designs).

After P&R has been finished, the configuration information is converted into a so-called bitstream file that can be directly programmed to the FPGA. FPGAs are programmed using special programming software (e.g., Impact for Xilinx FPGAs [Xil11d]), that connects to the programming interface on the FPGA using the industry standard JTAG interface. FPGAs can also configure themselves by reading the bitstream from an external memory (e.g., Flash) on power-up, which is important for SRAM-based FPGAs.

While the process to create the bitstream is highly automated and does only require manual attention for particular scenarios (e.g., defining the output pins or setting special operational conditions), programming in VHDL or Verilog is in fact a major hurdle for software developers to start working with hardware due to the fundamentally different concepts. For this reason, many groups invest efforts in increasing the development productivity by providing tools that support higher level programming (see Section 3.4). However, such tools currently can only generate high performance designs for particular applications. If high performance is required for an application that is not perfectly supported by one of these tools, handcrafting a proper design using VHDL or Verilog is inevitable.

2.1.3 Partial Reconfiguration

Partial reconfiguration (PR) is a technology that allows to reconfigure (currently inactive) parts of the resources of the FPGA while the remaining part keeps operating. This allows to build hardware, that can dynamically adjust to changing operating conditions. Examples are the type of current network traffic [Alb10], the selected transceiver model in a software-defined radio environment [He12], or the best fitting sort implementation for a large database [Koc11]. This mode of operation is also known as dynamic partial reconfiguration.

In this fashion, the chip can be smaller (and therefore cheaper), as currently unused logic does not allocate logic resources. Technically, DPR is supported on Xilinx Virtex FPGAs by an on-chip reconfiguration interface, the internal configuration access port (ICAP) [Xil12c], that understands partial bitstream data and has access to the entire configuration memory of the device. This interface can be accessed by internal logic or an embedded processor, so that the device can trigger reconfiguration activity autonomously of an external controller (known as autonomous DPR).

PR has been introduced by Xilinx for parts of their FPGA portfolio for several years now, but since the beginning only as experimental feature as part of an early access program with restricted participants. This is why PR and DPR solutions are especially prominent in the academic sector. Xilinx officially introduced partial reconfiguration just recently with the release 12 of the Xilinx FPGA programming software suite (ISE, [Fie10]).

To enable a design for PR on a Xilinx FPGA, a designer needs to reserve some portions of the FPGA resources for partial reconfiguration (so called partitions). The parts of the design that should be reconfigurable (e.g., different acceleration implementations) are then assigned to these partitions. This assures that during P&R no static logic will be placed inside these reconfigurable regions that is accidentally overwritten during reconfiguration. The corresponding P&R tool (PlanAhead for Xilinx) then generates bitstream files for the static logic and different partial bitstream files for the reconfigurable modules that do only reconfigure the FPGA resources inside the previously defined partitions.

One of the constraints when using partial reconfiguration is, that the reserved space for the partitions must be defined upfront and cannot be sized dynamically afterwards. Furthermore, if there are multiple partitions in a design, the resources of partitions cannot be shared between modules. The designer must set an upper limit, which can be a problem if modules are later developed that might be larger and do not fit into a single partition. To overcome this, there are some academic proposals that allow to arbitrarily mix reconfigurable modules inside partitions [Koc08, Alb10], but such setups are not supported by the official vendor tools due to side effects that come with such elastic setups (e.g., variations in delay times).

2.1.4 Special FPGA architectures

Beyond the conventional, broad-range FPGAs, where the market is dominated by the two vendors Xilinx and Altera, smaller start-ups begin to offer solutions that target especially high-performance applications in the communication and networking domain. Two of these are Achronix and Tabula.

2.1.4.1 Achronix Speedster

Achronix proposes their PicoPipe [Ach08] technology to achieve speeds of up to 1.5Ghz on an FPGA (e.g., a Virtex 5 has a maximum frequency of around 500 MHz [Xil12b]), which makes it ideal for high performance networking applications. The basic design of their Speedster FPGAs [Ach12b] is not that different from other FPGAs. A so-called PicoPipe logic fabric consists of reconfigurable logic blocks (RLB, contains LUTs, Flip-Flops and RAM) that are surrounded by switch blocks in each corner, The switch blocks route global signals through the fabric.

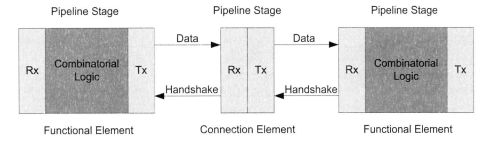

Figure 2.3: PicoPipe pipeline stages [Ach08]

What makes PicoPipe different from other technologies is the concept of data tokens [Ach08]. Generally, data tokens contain the logic value at a particular clock edge. PicoPipe uses data tokens with data and clock edge merged together, rather than acting in response to a clock cycle. The propagation of data tokens is supported by so-called connection elements (CE, see Figure 2.3). CEs are novel elements [Ach08] that can be initialized to be either state holding or not state holding. When they are state holding, they behave like conventional registers. In the other case, they behave like a repeater. However, while on the first look this seems like a simple wire, each of the pipeline stages implemented by non-state holding CEs can contain a different data token.

One major advantage of that concept is that pipeline stages can be implemented transparently without affecting the logic operation [Ach08]. While in conventional designs each pipeline stage automatically adds an additional cycle, this is not the case for PicoPipe. Only a small delay is added. The possibility to insert such transparent pipeline stages anywhere in the data path has a major impact on performance, as the heavy use of pipelining assures that data only has to take short paths while it is processed. This increases the maximum operating speed and with that the throughput. As this is especially useful for 10G networking applications, Achronix provides a development platform targeted for that domain that consists of a development board equipped with embedded 10G interfaces and a speedster FPGA [Ach12a].

2.1.4.2 Tabula ABAX

Tabula is an FPGA start-up whose vision is to increase the density of logic by making heavy use of DPR to create a virtual device with multiple stacked layers, that from the outside looks like a large three-dimensional single chip. The technology behind that vision is called Spacetime [Tab10]. Spacetime uses dynamic partial reconfiguration with multiple Ghz (in comparison, a current Virtex 5 FPGA has a reconfiguration speed of 100 MHz) to reconfigure parts of the device on-the-fly while data is processed.

This process is transparent to the user. Spacetime devices provide LUT-equivalent logic units (called MegaLUT) and multiple external I/Os as any other FPGA. An automatic compiler maps standard designs into Spacetime designs that can be run on Spacetime-enabled devices, such as the Tabula ABAX. The high logic capacity in combination with the high processing speed makes ABAX devices an interesting candidate for high-speed networking applications.

In addition to offering ABAX FPGAs, Tabula provides networking related prototype and production boards, such as the 10G NetASAPdc [Tab11], that is especially targeting applications with demands for associative and exact matches (switches, routers, firewalls).

2.2 Hardware Development Boards

Hardware for FPGA-based 10G networking is offered by multiple vendors. The following section describes two of them in more detail, that have been used as part of this thesis.

2.2.1 BeeCube BEE3

The BeeCube BEE3 [Dav09] is a quad-FPGA platform that has been designed for compute intensive applications, but due to their multiple 10G Ethernet ports does also fit very well for network application development.

Figure 2.4: BEE3 reconfigurable computing platform [Bee09]

The BEE3 comes in different configurations, dependent on the target scenario. The BEE3 model used here has been equipped with four Virtex 5 FPGAs, 2x LX155T, which is a general purpose FPGA with a medium number of logic resources and 2x SX95T, which especially contains a high number of DSPs for compute applications and less logic resources. Figure 2.5 shows the schematic overview of the architecture.

Each FPGA has a connection to two dedicated 10G CX4 Ethernet ports. Therefore, one can select any of the FPGAs for network connectivity. Furthermore, each FPGA has a connection to a dedicated memory bank that is in the current configuration equipped with 2 GB of DDR2 memory. Additionally, the FPGAs themselves are connected in a ring-structure, so that communication between the FPGAs can be established. In addition to the network connectivity, each FPGA has connection to an 8-line PCI Express interface, through which it can be connected to a host PC for high-speed data transfer.

The board contains standard JTAG programming connectors, so that development on the BEE3 can be made using the standard Xilinx tools. However, there is no particular software support for networking applications, so the designer has to develop everything from scratch here.

The BEE3 is housed in a separate box and can be easily mounted in a rack. But actually, the use of the relatively old and slow DDR2 memory is a problem for data intensive networking applications, as DDR2 memory is not fast enough to hold buffers in external memory in that case.

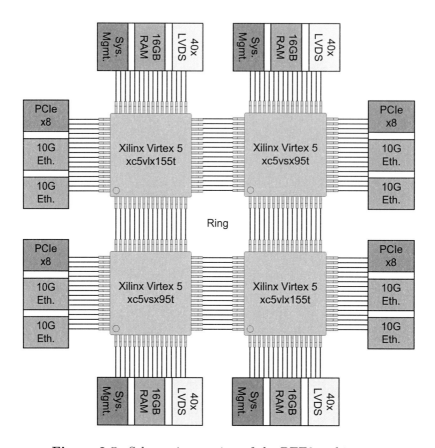

Figure 2.5: Schematic overview of the BEE3 architecture

2.2.2 NetFPGA 10G

NetFPGA 10G [Net11b] (see Figure 2.6) is the successor of the very popular NetFPGA 1G card. The NetFPGA 10G cards have been developed by the NetFPGA project (a collaboration between Xilinx, the University of Stanford and the University of Cambridge) to provide an open source environment for FPGA-based network development. They have been made available at the end of 2010.

For academic use, the hardware for the NetFPGA card is sponsored by several chip vendors. The card contains a Virtex 5 LX240T FPGA, one of the largest Virtex 5 series FPGA, and four 10G SFP+ transceiver slots, which can be equipped with fiber and copper transceivers to support different network environments.

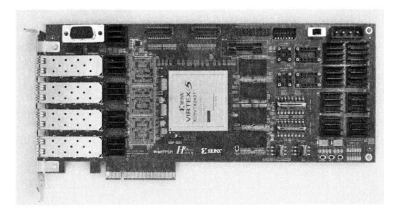

Figure 2.6: NetFPGA 10G FPGA hardware board for networking [Net11b]

As said, an important factor for networking applications is the availability of high-speed external memory. Fortunately, NetFPGA 10G is well-equipped here. It provides three dedicated QDR SRAMs, each with 9 MB of memory, that are well suited to hold short-term buffers and 288 MB of RLDRAM II, which provides low-latency access to larger portions of data. Thanks to its PCIe x8 connector, the NetFPGA card can be easily plugged into any server and connected with software running on the host PC (if required).

Beyond the hardware board, the NetFPGA project does also provide a rich set of application components and development tools that aid in the design of hardware-based networking applications. However, the codebase for the 10G version is still under development and has been just recently made available for public beta [Net12a].

3 Prior and Related Work

This Chapter discusses prior and related work matching the three main areas introduced in the first Chapter: Hardware Support for Network Security, Platforms for Hardware-Based Networking and High-Level Compilation. Furthermore, honeypots will be discussed in a separate Section due to the relevance for the application example.

3.1 Hardware Support for Network Security

The use of FPGAs to accelerate network security applications has been a popular field of study in recent years. In a recent survey on hardware support for network infrastructure security, Chen et al. [Che11] separate the research efforts in four different classes: Packet Classification, Pattern Matching, TCP Stream Processing, and Internet Worms and Distributed Denial of Service (DDoS) Attack Detection and Containment.

Packet classification (focused on evaluating header data) and pattern matching (uses regular expressions for scanning the payload of packets) are significant base classes, while the other two can be also rather seen as applications than separate groups. Internet worms and DDoS attack detection is one of the example applications of the category anomaly detection. TCP stream processing does support the other categories by providing, e.g., stream reassembly for better inspection of communication over TCP channels (e.g., if malicious data is spread across multiple fragments).

In alignment with the focus of this work, five categories have been finally defined to group related work, that extend the classification of [Che11]: Packet Classification, Packet Matching, Anomaly Detection, High-Level Hardware-Based Applications and Communication Support.

A common application for **packet classification** in the security domain is a packet filter. Based on certain rules, a packet filter decides if a packet is dropped or forwarded (e.g., implemented in a network firewall). Packet classification and corresponding algorithms have been addressed multiple times for FPGA implementations, sometimes especially for packet filtering [Jed08], but also in a more general context [Nao08a], as such high-speed classification operations are also especially relevant in routers and switches.

Pattern matching is the foundation for many rule-based IDSs, e.g., Snort [Roe99]. Pattern matching is closely related to regular expression matching, as patterns in the network infrastructure security context are often given as a regular expression. Significant research effort has been invested in improving regular expression matching throughput for such systems using FPGAs [Sid01, Yam08, Wan11, Yan12]. For evaluating the performance of such pattern matching implementations, the rule-base of the popular Snort open-source IDS is commonly used, as Snort is the de-facto standard in this area.

In contrast, an **anomaly-based IDS** does not scan for specific patterns in the network traffic. Instead, it establishes a statistical baseline for normal network operation and compares current traffic statistics against this baseline to detect anomalous behavior that could point to a malicious activity. An example use case is worm spread detection on global network routers [Wag05]. However, raw hardware implementations are rather seldom in this area. This is because algorithm complexity can get very difficult and requires to store high volumes of statistical data [Pat07]. Actually, hardware acceleration is therefore mainly employed for collecting the network statistics at real time [Bar08].

While the previous three categories cover rather passive devices, the implementation of **active communication endpoints** entirely on FPGAs has been rarely addressed. Published work does propose to offload some work to hardware accelerators [Che06a, Sad11], but not the entire communication. Currently, the research on such FPGA-based Internet communication applications is just at the beginning (see also [Blo12]).

A basic element for such applications is the **implementation of the basic Internet communication protocols** (IP, UDP, TCP etc.). Due to the complexity of TCP, past work in that area does often address only the acceleration of a subset of the protocols, e.g., compute intensive checksum calculation, combined with a generic processor that manages the rest of the protocol (called offloading) [Jan09]. Just recently, especially commercial companies started to offer entire protocol stacks that support autonomous communication for line rates of 10 Gbit/s [Int12b, PLD12] on FPGAs for custom application design, but on the academic side this still has been addressed very rarely and not for 10 Gbit/s.

The following subsections now discuss the different topics in more detail.

3.1.1 Packet Classification

In [Loi07], Loinig et al. propose an FPGA-based packet filter that achieves gigabit line rates using a simple linear search as matching algorithm. They propose this algorithm as being best suited for FPGA implementations due to the possibility to make use of the internal high speed memory blocks. With a rule database of 32 entries, their system achieves a throughput of 1 Gbit/s with multiple parallel search engines running on an FPGA with 125 MHz. While this might be an option for very small rule sets and moderate speeds, such a solution does not scale for 10 Gbit/s or more.

One technology component that is often used in this context (either solely or in combination with other algorithms) is a ternary content addressable memory (TCAM) [Yu04]. A TCAM is based on the content addressable memory (CAM), which is a special memory architecture that allows searching for the content inside a memory within a single operation instead of requiring to scan through all addresses manually. E.g., in a simple case this can be used to classify packets based on a short list of destination IP addresses. However, network addressing knows the concept of netmasks, which does not require an exact match, but only a match for parts of the data word. This is the domain of TCAMs. A TCAM allows matching 'X' (means anything) as third state beyond '1' and '0'. In this manner, matching a subnet address can be easily implemented.

As inherited from the CAMs, the advantage of TCAMs is their enormous search performance. Developments in chip architectures and manufacturing processes nowadays allow creating huge TCAMs at reasonable prices. That is why nearly every hardware-accelerated commercial router, switch or firewall-solution relies on TCAMs for certain tasks. E.g., CISCO uses it for its routers in the nexus 7000 series [Cis11a], which makes heavy use of TCAMs to implement access list based filtering directly on the interface modules. The NLA9000 content processor family from Netlogic (aquired by Broadcom [Bro12]) uses TCAMs to classify up to 400 million packets per second with a rule capability of more than 100k entries [Vai11].

However, while TCAMs are very flexible devices they have a higher power consumption and require much more chip area [Vai11] in contrast to regular memory, which prevents easy scaling of TCAM-based architectures. For large and more complex rule databases (TCAMs are not so efficient for range-based searches [Son05]), other algorithms might be better suited to achieve high speed rates.

A prominent choice is to use decision-tree based algorithms, as they offer more flexibility in terms of rule definition. Based on the rule database, a custom decision tree is generated. Keys (generated from the packet headers) are then used to traverse the tree. The combined BV-TCAM architecture from Song and Lockwood [Son05] achieves a data throughput of 2.5 Gbit/s on a rule database of 222 entries (stored in on-chip BlockRAM) using a combination of tree-based algorithm (for ranges) and a TCAM (for single value lookups). However, the selected bit vector algorithm does not scale with the number of rules, as the space required to combine bit vectors corresponds to the square of rules [Vai11].

Vaish et al. [Vai11] proposed another solution that is based on the EffiCuts algorithm. EffiCuts tries to achieve memory reduction by generating multiple decision trees. However, instead of doing a raw hardware implementation, they implemented the algorithm for the PLUG processor [DC09], which is a hardware-accelerated processing platform for research on lookup algorithms in routers. The EffiCuts PLUG processor achieves a throughput of 45.4 GBit/s regardless of the number of rules. In this fashion, the system scales better than FPGA-based solutions and updates are easier. However, native dedicated hardware implementations still outperform such programmable solutions in terms of performance. In 2010, Qi [Qi10] proposed a solution that uses HyperSplit, a memory-efficient tree search algorithm to achieve a throughput of 142 Gbit/s for 50k rules implemented on a Virtex 6 FPGA.

The importance of hardware acceleration in that area has also an impact on commodity hardware in that sector. E.g., the Intel 82599 10G network interface chip provides a hardware filter section with up to 32.000 rules [Int12a]. A performance evaluation showed the power of these hardware filters for small packet sizes. Processing 64B packets generates a load of 100% on a standard CPU (6-core Xeon), while the same packets are processed with a load of zero in a server equipped with such a network card [Der10].

In addition to filtering, another usecase for packet classification can be the collection of network statistics (e.g., traffic rate for a particular service), that in turn can then be used as the base for anomaly-based IDSs to perform further analysis (see also Section 3.1.3). Note that research on classification and lookup algorithms is also heavily employed for other parts of networking equipment, e.g., Open Flow switch implementations [Nao08a] or large-scale IP address lookup [Le10] for the use in backbone and core routers.

3.1.2 Pattern Matching

In the majority of cases, a regular expression will be implemented using either a Deterministic Finite Automata (DFA) or a Non-Deterministic Finite Automata (NFA) [Hop01].

DFA implementations consist of a combined transition table that contains all state transitions for all regular expressions that should be matched, while NFAs implement separate state machines running in parallel. The NFA approach does well match the parallel design of FPGAs, as each state machine can be implemented as dedicated hardware. Based on the work of Sidhu and Prasanna [Sid01], in 2002 Hutchings and Franklin [Hut02] already demonstrated performance gains of up to two orders of magnitude over competing software solutions at that time.

Nowadays, NFA-based FPGA implementations of large scale regular expression matching currently achieve data rates of 10 Gbit/s or more for comparing thousands of rules [Yan12, Wan10, Gan10, Yam08, Yan08]. Techniques, such as prefix sharing [Bis06], centralized character encoding [Bis06], and relation similarity [Kos11] are used to reduce the total number of FSMs. Multi-character input [Yam08] and pipelining [Hut02] improve the overall efficiency.

In addition to the hardware architecture, these solutions also propose compilers that automatically translate rules (e.g., Snort rules) into an FPGA implementation. Otherwise the complexity of the FSMs would not be manageable. Furthermore, some work also combines packet inspection and filtering [Kat07].

But even if there are automatic compilers, raw hardware architectures have a disadvantage for frequent updates due to the long FPGA design compilation process (see Section 2.1.2). This in turn is one advantage of DFAs. Due to the table-based approach, dynamic updates of systems can be performed easily. Furthermore, storing the transition states in a table does not consume much logic resources on the FPGA that needs to be synthesized. However, an often mentioned drawback of DFAs is, that tracking concurrent matching paths for multiple regular expressions could lead to massive state explosion [Wan10]. Several research groups work on the DFA state explosion issue by using compression techniques [Qi11, Wan11] to reduce the DFA state table complexity. On a recent Virtex 6 FPGA, Qi et al. [Qi11] achieve a speed of up to 10 Gbit/s for the entire Snort rule set.

Korenek et al. [Kor10] propose a solution that aims at combining the advantages of both worlds by dividing the regular expression set between NFA and DFA implementation based on collision detection, which reduces the logic resources required to implement the NFAs on the FPGA, while avoiding state explosion in the DFA table. A similar approach is also employed by the commercial T2000/T2500 content scanning processor from Tarari [LSI10], that supports up to 20 Gbit/s programmable regular expression parsing.

In the past, also CAMs and TCAMs have been employed for pattern matching [Yu04, Sou04]. However, due to the previously mentioned drawbacks of TCAMs (see Section 3.1.1) researchers nowadays focus on the DFA and NFA approaches.

3.1.3 Anomaly Detection

An issue with anomaly-based IDSs in high-speed environments is that software-based solutions might not be capable of collecting real-time network statistics without packet loss. While sampling might be a possible solution for certain applications, accurate systems need to capture every packet flying by. This is a classical domain for hardware acceleration. As the Cisco NetFlow standard [Cis11b] has been established as a common format for collecting traffic statistics (network flows etc.), hardware solutions mostly aim at supporting the collection of NetFlow-compatible flow data.

The FPGA-based NetFlow probe from Bartoš et al. [Bar08] is able to monitor flow data at the speed of 3.2 Gbit/s. A PCI Express bus connection is used to transfer the data to the host for later analysis. This is often the case for such anomaly-based solutions. While hardware can assist in the collection of flow data, processing the data is often done offline due to the complex algorithms that can be employed here (e.g., machine learning [Wag11]). This work has been continuously improved and is now sold as a product called FlowMon from INVEA-TECH, achieving full 10 Gbit/s line rate [IT12].

Hardware-accelerated anomaly detection and especially the collection of flow data has been also extensively studied by Vojtech Krmicek in his PhD thesis [Krm11]. His work supports high-speed real-time collection of network statistics using the NetFlow standard extended with new functionality to support bidirectional flows. This allows to distinguish between requests and responses, which in turn allows to classify network components into servers and clients for higher accuracy when detecting attacks to particular servers. Krimcek uses custom hardware for this task, that has been developed based on the INVEA-TECH COMBO hardware platform [IT10].

A typical use-case for the collected NetFlow data is worm outbreak detection. Wagner et al. [Wag05] calculate compression ratios using NetFlow data on a host PC and can detect a worm outbreak by changing ratios. However, there are some proposals that avoid post-processing on a CPU. E.g., [Fae09] propose a system that detects worm outbreaks by counting the repetitive occurrence of strings in network packet data. The system has been entirely implemented on an FPGA to support gigabit line rate speeds. Mathematical calculations, such as the probability density function for the counters that are used to build a model of the network, are implemented directly on the FPGA. Thresholds are then set to define the alarm levels.

Another example for a hardware-only anomaly-based detection unit is the work from Chen [Che08]. He focuses on the detection of DDoS attacks by calculating the power spectral density of the sampled network traffic. Again, the FPGA is accelerating the computation of the algorithm.

As an alternative to NetFlow collectors, Kobiersky et al. [Kob09] provide a solution for programmable header field extraction, that allows to define headers to be extracted using an XML description. A compiler then generates a VHDL module with a state machine that performs the actual header extraction process. Their system achieves data rates of around 10 Gbit/s on a Virtex 5 FPGA. But in contrast to, e.g., FlowMon [IT12], this system cannot be updated with dynamic rules.

[Pan12a] proposes a hierarchical flow table to identify applications based on high volume traffic flows. As collecting such data is especially a memory intensive process (an entry needs to be stored for every open network session), they store millions of flows by using external DRAM as base storage and an on-chip cache implemented in BRAM that provides fast lookups for active flows. To increase the cache hit rate, they further propose a so-called Adaptive Least Frequently Evicted replacement strategy, where long-running flows with higher activity are preferred. Results show a throughput of 28 Gbit/s for the smallest 40B packets.

While anomaly-based IDSs with their capability to detect previously unknown attacks (as long as these attacks lead to deviations in the observed behavior) offer interesting capabilities in addition to rule-based system, one of their major drawbacks are false-positive alarms. These become a significant problem in large-scale deployments of such systems: A human expert has to evaluate the individual "alarms" indicated by the anomaly-based IDS for their actual significance, which actually hinders broad usage in commercial environments and is another important research topic next to raw performance (see, e.g., [Vig09]).

3.1.4 High-Level Applications

In the previous three Sections, the focus was on the packet layer, where systems inspect network data but do not actively take part in a communication. In contrast, this section covers work that implements active network endpoints on dedicated hardware. Such applications are named "high-level applications" in this thesis.

Actually, transferring an entire communication application to an FPGA has only been very rarely addressed. While first mentioned by Fallside in 2000 [Fal00], there has not been high activity in this field. Most of the work does only transfer compute-intensive tasks to the FPGA and use embedded GPPs for complex protocol processing. E.g., Gonzalez proposed a SSH accelerator [Gon05] for a Microblaze CPU. Sadoun presented a DNS server implementation [Sad11] on the Virtex 5 FPGA that contains embedded acceleration for DNSSEC. In both cases, especially the cryptographic operations have been implemented on hardware.

Outside the area of network security, a hot topic is the implementation of high-frequency trading (HFT) applications directly on the FPGA without any processor that eats up valuable cycles [Loc12]. These applications aim for the lowest possible latency to be the first performing a transactions and so gain financial benefit over other trading applications. Leber et al. [Leb11] propose a solution using a DSL in combination with a micro-programmable hardware engine to support trading applications on the FPGA. They achieved a four times lower latency than compared with a state of the art server.

3.1.5 Communication Stacks

Indispensable for the implementation of higher-level network communication on FPGAs is the availability of an efficient implementation of the basic Internet protocol stack [Bra89]. In that context, especially the transmission control protocol (TCP) [Pos81c] needs particular attention, as it is the preferred protocol for common applications, such as web, email and other services on the Internet (a recent evaluation shows that around 90% of Internet traffic uses TCP [Hur11]). Furthermore, it has a higher complexity than other protocols of the stack (e.g., UDP or IP), so that the hardware TCP implementation of such a communication stack does often define the limit for the overall performance. Due to the relevance of TCP, such communication core implementations are sometimes also simply named hardware TCP/IP cores, even as they may contain other relevant protocols of the stack (such as ARP or UDP) as well.

Actually, implementing such a complete Internet communication stack on FPGAs has only rarely been addressed in the research community, especially for 10G high-speed environments. Furthermore, in the majority of cases earlier work covers only gigabit networking, as proper development hardware for 10G environments (e.g., the NetFPGA 10G card [Net11b]) has become available for a broad range of users just recently.

Schuehler and Lockwood [Sch02, Sch04] began focusing on FPGA-based TCP processing for gigabit line rates in 2002. But their intention was to monitor TCP streams in switches and routers on their way between sender and receiver instead of enabling endpoint communication. They motivate this with the sheer complexity of an entire TCP endpoint implementation on hardware. Later work did also focus on TCP stream reassembly [Vu11, Yua10], which is important for building monitoring devices or modern routers that route packets based on flows, but does not support an active communication. Yuan et al. [Yua10] target 10G networks, but they only simulate their design and did not provide implementation results.

Dollas et al. [Dol05] were one of the first that aim at providing an entire communication stack, but their solution (achieving around 350 Mbit/s) does not reach the performance of high-speed networks and the project was not continued. In 2009, the Fraunhofer HHI [Gre09] presented a feature-complete 10G TCP core for FPGAs as a product, but unfortunately at the beginning it was able to serve a single concurrent connection only, which highly restricts its use. However, in 2011 a new version has been released that is now able to handle more than a single connection.

As an intermediate solution, TCP offload engines have been proposed for a couple of years that support CPUs with compute-intensive protocol processing (e.g., checksum calculation) for high-speed operation, but leave complex parts of the protocol (e.g., flow control) up to the CPU [Jan09]. This area has also been addressed by commercial companies that focus on providing such 10G TOE engines to embedded systems in exploiting the full performance of a 10G network link [Myr12, Che06b]. One application scenario that drove this development was the acceleration of iSCSI communication, a protocol that implements access to storage devices over IP networks and therefore requires low-latency and high throughput [Che06a].

However, in the past two years, activity on the commercial side in the field of full-features communication cores has ramped up. With the upcoming of more powerful FPGAs and corresponding hardware platforms, and supposedly driven by the exploding interest in FPGA-based high-frequency trading applications [Leb11], a couple of companies presented or announced [Int12b, Gro12a, PLD12] complete TCP/IP stacks for the use in FPGAs at data rates of 10 Gbit/s or more. Actually, these cores support various configuration parameters (e.g., number of supported connections, with or without fragmentation support) to allow the designer to adapt the core to the particular needs (functionality vs. size vs. speed) [PLD12], as various functionalities in TCP (e.g., flow control, fragmentation) are disadvantageous for a hardware implementation.

While not focusing on TCP, but still a very interesting work to mention here is the TLS/SSL accelerator presented by Isobe et al. [Iso10], as TLS/SSL is one of the most prominent protocols for transport layer encryption (e.g., for secure web pages) on the Internet. They designed a hardware-based TLS/SSL accelerator that implements the entire stack on hardware. Their design includes protocol processing (a minimal set of TCP functionality plus SSL/TLS), cryptographic processing (RSA, AES, SHA1, RC4, MD5) and encrypted data exchange. The FPGA prototype achieves a throughput of 10 Gbit/s, while supporting 12.000 key exchanges per second. While the number of key exchanges per second is lower than compared to a solution based on a CPU together with an accelerator, the overall throughput and power consumption is much better. Furthermore, such a solution allows building hardware-only applications with encryption support.

3.2 Hardware-Based Network Platforms

The following Section lists common platforms and proposals related to networking, that support designers with functionality for building entire applications with hardware support. In that context, three different categories are highlighted: FPGA development platforms, network processors and custom architectures. FPGA development platforms provide an environment that allows rapid creation of dedicated hardware networking applications on FPGAs. Network processors allow to program accelerated networking applications, which make use of hardware accelerators for certain tasks and a general purpose processor with networking extensions for other work. Custom architectures cover some work that proposes special architectures, which support networking on hardware.

3.2.1 FPGA Development Platforms

In the area of FPGA-based networking, doubtlessly the most popular research platform is NetFPGA [Loc07]. NetFPGA has been developed by a research group at Stanford University and presented to the public in 2007. The first version of the hardware board (called NetFPGA 1G) consists of a Xilinx Virtex 2 Pro FPGA, four 1G network interfaces, external SRAM and DDR2-SDRAM as well as a PCI interface connector. The interesting part of NetFPGA is that it does not only provide a raw hardware board, but also a rich set of software tools. It provides an extensive set of sample projects and a complete simulation and build environment that allows researchers to quickly start working with the platform. Furthermore, the hardware is sold to academic institutions at a discounted price.

One of the reference projects is a flexible router [Nao08b], that demonstrates the basic concept of the platform. Packets go through a main data path, where modules can be implemented that work on the packets. Figure 3.1 shows the basic core architecture of the NetFPGA platform.

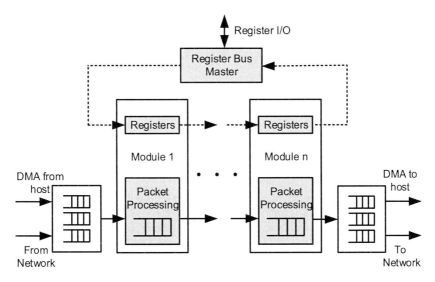

Figure 3.1: NetFPGA core architecture [Nao08b]

All the four external network ports can be connected to the main data path, while an internal header added to every packet maintains the mapping information. Furthermore, the PCI interface can be used to connect the system to a host PC for management operations. This reference project does also show the intention for what NetFPGA has been created for: to provide a suitable environment to use FPGA-based networking for switching and routing in teaching and classrooms [Gib08].

The attractiveness of the project led to a widespread use of this platform and other groups started to contribute code. Currently, the NetFPGA 1G project table lists over 50 different projects [Net12b], e.g., different switching and routing implementations. In 2011, the NetFPGA team released the next generation of the NetFPGA card (called NetFPGA 10G), that now contains four 10G network interfaces and a more powerful FPGA. Details on this card have been already given in Chapter 2.2.2. However, at the beginning the 10G codebase was still in beta as converting the NetFPGA core applications to 10G has turned out to be more complex than expected.

A commercial platform that has a comparable focus as NetFPGA is NetCope [Mar08] from INVEA-TECH. NetCope also consists of a FPGA hardware and software part. However, NetCope primarily targets commercial research and production use, e.g., for telecommunication providers and was bound to the INVEA-TECH Combo Boards [IT10]. However, with the release of the NetFPGA 10 card they presented a port of NetCope to be run on the NetFPGA 10G and they announced to provide special academic licensing for this software package [Kor11]. This might be interesting for future developments.

They started to offer 10G solutions already in 2009 with their Combo V2 and LXT cards [IT10], that contain two 10G interfaces, QDR SRAM and DRR2 SDRAM, a Virtex 5 FPGA and a PCI express connector. INVEA-TECH furthermore sells solutions based on their platform. One of these products is a FlowMon probe [IT12] (see also Section 3.1.3), which is able to monitor high-volume network traffic and to generate network statistics for later analysis.

3.2.2 Network Processors

Network processors are special processors that are designed for high-speed packet processing. They offer the flexibility of a software-programmable environment, but achieve a high performance compared to a processor-only solution due to dedicated processing elements for packet processing. Basically, there are two architectural principles for network processors [Pey03]. One is single instruction - multiple data (SIMD, see Figure 3.2-a): multiple processing elements (cores), that support the same instructions, process packets in parallel. A task scheduler manages the allocation of packets to processing elements. The other one is pipelining (see Figure 3.2-b): multiple cores, where each core performs only a particular operation, are arranged in a pipeline and a packet is forwarded through a set of cores for complete processing. This mode of operation provides better efficiency, as the cores can be better tailored to the tasks.

Commercial network processors are offered from various vendors (Intel, Marvell, Broadcom) targeting speeds of 10 Gbit/s and more in different environments. E.g., the Marvell Xcelerated and the Broadcom Trident+ product family covers high-speed switches and routers [Mar13, Bro10] by supporting high-speed lookup operations. Often network processors are also combined with dedicated hardware accelerators for specific tasks, such as checksum calculation, encryption or even

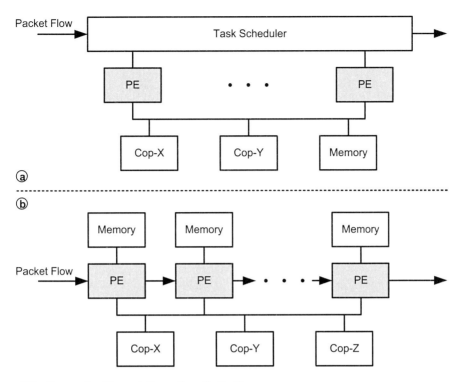

PE = Processing Element; Cop = Specific Co-Processor

Figure 3.2: Basic operating modes of network processors [Pey03]

pattern matching. The IXP2800 family from Intel [Int04] is a very popular general network processor that has an ARM-based RISC architecture and was the first that contained embedded accelerators for encryption.

One of the advantages of network processors is, that they often support an integrated development and debugging environment. However, each network processor has a specific instruction set, which on the other hands makes it difficult to switch between products. Furthermore, even if pipelines can be flexibly structured, network processors still have shared resources (e.g., memory) that limit the achievable performance (see also [Sir08]).

Another issue is, that software-programmable processors are vulnerable to compromising attacks, as they can execute any code. This is especially dangerous, if the data path in such a networking application is implemented using software. This issue has been addressed by Chasaki et al. [Cha10] with their proposal for a secure network packet processor that uses runtime monitoring techniques to detect the impact of a compromising attempt (e.g., abnormal program paths taken by the processor).

They implemented a prototype of their processor on the NetFPGA platform. The monitoring functionality has been implemented directly in hardware, so that it cannot be altered. In this manner, attackers can still change the program of the processor, but this change will be detected and countermeasures can be taken (e.g., drop the packet and restore the program). They show a performance overhead of roughly 5% for their monitoring extension, but actually this number is based on an average throughput of around 65 Mbit/s, so that scaling the approach to 10 Gbit/s or more is questionable.

3.3 Custom architectures

Beyond such broad-range FPGA networking platforms and traditional network processors, there are several proposals for custom architectures targeting networking hardware, such as custom pipeline structures [Kar10] or reconfigurable network-on-chips [Alb06].

[Kar10] propose an FPGA-based network processor for multi protocol label switching (MPLS), a technique that is used in carrier-grade backbone Ethernet environments. One of the key parts of the architectures is a custom pipeline structure that has been designed to overcome the shortcoming of traditional pipelines, where packets need to go through the entire pipeline even if processing in later stages is not required.

Each label is processed by a so-called mini-pipeline that contains the entire steps for a single label. The FPGA contains multiple of these mini-pipelines for all labels that could exist. Whenever a label has been processed, a dispatcher checks whether the packet is done or if there is another label that needs to be processed. Their design implemented on a Virtex 5 LX330T is targeted for a throughput of 100 Gbit/s, but it has to be said that they split header and payload of a packet and do only process the header information with their architecture, which significantly reduces the data path requirements. Packets are later reassembled when send out of the system.

A similar idea is followed by the FlexPath NP project [Ohl10]. Instead of always taking a single path, an application-dependent path decision is made upfront to packet processing. Figure 3.3 outlines this concept.

For performance reasons, the path dispatcher is implemented as dedicated

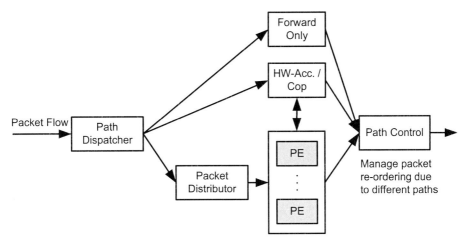

PE = Processing Element; Cop = Specific Co-Processor

Figure 3.3: FlexPath NP core architecture [Ohl10]

hardware. Paths can then consist of CPU resources as well as hardware accelerators. They implemented a prototype on a Virtex 4 FPGA that was designed for a speed of 3.2 Gbit/s. In contrast to [Kar10], Flexpath NP is more generic as it handles the entire packet.

DynaCORE [Alb06, Alb10] is a system that makes heavy use of FPGA partial reconfiguration to support high-flexible acceleration. DynaCORE does not provide any network-related specific functionality, but instead a basic infrastructure for connecting reconfigurable hardware accelerators to a generic network processor. They use DPR in an extensive way that goes far beyond the partition-based reconfiguration proposed as the standard mechanism by the vendors (see also Section 2.1.3).

DynaCORE defines reconfigurable regions that can be used by multiple application acceleration modules (e.g., packet encryption) in different configurations. E.g., a region can host one large or two smaller modules. The interconnect between these dynamic modules is provided by a reconfigurable network on chip, called CoNoChi. Instead of providing a fixed switch setup, CoNoChi switches are reconfigurable modules just as the application accelerators. Whenever a reconfiguration of a module occurs, switches will be reconfigured as well to provide the required connectivity at the requested bandwidth. Packets (messages) are routed using a specific CoNoChi header. The concept has been demonstrated by implementing it on a Virtex 4 FPGA, achieving gigabit link speeds between the accelerators.

3.4 High-Level Hardware Compilation

As the complexity of FPGA programming (see also Section 2.1.2) is one of the key points that prevent widespread use of FPGAs in research and commercial environments (there are in fact many more software developers than hardware engineers), there is a high activity in industry and academia to lower the barrier by providing better programming solutions on a higher level.

Basically, there are two major approaches for high-level hardware compilation: Using well-known high-level general programming languages (e.g., C) or special domain-specific languages as source. While general programming languages have the advantage that they can express any problem, domain specific languages score with their focus on a particular problem, where they reach a higher programming efficiency. Brebner [Bre09] shows a productivity gain of more than six times for using a domain-specific language such as G versus a similar implementation in hardware-synthesizable C.

Beyond the programming aspect, also the hardware generation part is of particular interest, as design compilation still can take hours or days for the mapping and place & route processes to finish, even if only a few lines of code have been changed. While the majority of tools target dedicated hardware generation, there are some projects that propose to compile micro-programmable hardware structures that are flexible up to a certain extent. While reducing the flexibility and performance by some points, this allows to perform small updates quickly.

3.4.1 General Purpose Languages

As general purpose languages are well known to software developers that already write the software part of an embedded application (e.g., consisting of a CPU and an FPGA), using such languages also for automatically compiling suitable hardware is one of the key objectives in industry and academia. Currently, C and C-derivates (e.g., System C) are dominant here, due to their relevance in the embedded system domain. Research projects, such as C-to-Verilog [Rot08] or Comrade [Gad07] continuously improve the capabilities of their compilers (loop unrolling, scheduling, memory management etc.).

In addition to the research groups, that work on that topic for years now, especially commercial companies have started a real product offensive during the past two years in that sector. An evaluation of a set of current high-level compilers is given by Meeus et al. in [Mee12]. Differentiation criteria between the single products are, e.g., the support for data types (fixed or float), local memory and C language constructs (pointers etc.). Furthermore, Meeus et al. used a Sobel edge detector as test application to verify the functionality of the different tools. Interestingly, the well-known hardware design tool providers Xilinx, Mentor and Synopsys acquired the technology for their high-level synthesis products from smaller companies. Obviously, they have been caught off guard by the huge market demand in such solutions.

Vivado HLS [Xil12d] by Xilinx is a generic C to HDL (VHDL, Verilog) compiler that has a broad range of functionalities. The compiler scans the code for control structures and converts them to processes and conditions. The data part is then modeled by, e.g., loop unrolling and proper scheduling. Vivado HLS supports various module interfaces (e.g., simple FIFO and the AXI protocol), which simplifies the integration with other FPGA logic. In their tests, Meeus et al. found the compilation process and the results most convenient amongst all their candidates. Xilinx integrates Vivado HLS into the Vivado suite of Xilinx development tools, which provides a seamless integration of hardware and software part. Actually, Vivado HLS is an advancement of AutoPilot from AutoESL, which has been acquired by Xilinx.

CatapultC [Cal12] by Mentor Graphics (bought from Calypto Design Systems), is another generic C to HDL compiler with a rich feature set. It supports, e.g., loop unrolling, creation of pipelines and data can be stored in registers or RAM. However, the Sobel edge detector test required some manual changes to achieve better performance (local buffering has not been enabled automatically), so that a background in digital system design is still required.

Synphony C [Syn12] by Synopsys (originally known as Pico, developed by Synfora) supports a subset of the C functions, but code generation requires parameters to be set as pragmas inside the code to achieve a high performance, so that hardware knowledge is required also here.

According to Meeus et. al., all the tools achieve good results, especially for compute-intensive applications that benefit from loop-unrolling and pipelining. From the usability aspect, Vivado HLS seems to be the easiest one to start with, if not being a hardware specialist. However, these generic compilers always have to deal with limited transferability of control structures into efficient hardware. Other work therefore proposes to use such compilers in combination with a GPP and to split the work between the GPP and the hardware accelerator [Koc10] to achieve ultimate performance.

As an alternative to the C-based solutions, BlueSpec [Blu10] has been proposed recently as a hardware language with a higher abstraction level. BlueSpec descriptions are written in BlueSpec System Verilog (BSV), a language based on Verilog. Functionality is described using so called Guarded Atomic Actions that define conditional operations for a particular hardware module. Clock, scheduling and dependencies are then handled implicitly by the BlueSpec compiler. The advantage of BlueSpec is that due to its HDL roots, it is easier to write compiler-compliant code than when using C.

3.4.2 Domain-Specific Languages

In the domain of hardware-based networking, domain-specific languages are a wide research field for various application scenarios. The following lists some of the current work.

The language G has been designed in particular for packet-header processing on FPGAs. It allows for flexibly specifying the packet format (fields and positions) and conditional rules for modifying these fields depending on packet contents. The programs are then compiled into hardware units, that can be run on the NetFPGA 1G card [Bre09]. While capable of payload processing, G lacks facilities such as regular expression handling and extended support for protocols above the level of processing individual incoming packets (e.g., TCP connections, generation of new packets for protocol interactions). The focus on header processing in G is also emphasized by the example applications, which deal with switching or MPLS routing [Bre09].

ReClick [Unn11] is a domain-specific language that allows describing packet operations based on basic field operations. These field operations are: get, set, insert and remove. ReClick components can then be described using these basic operations and are then compiled into dedicated hardware components. The architecture of ReClick components consist of tables that control the field operations defined by the component specification. A scheduling module on the output of each component selects the next component for each packet (based on basic rules defined in the ReClick description). However, complex modules, e.g., an IP lookup, need to be developed manually. The authors of [Unn11] implemented a IPv4 router for the NetFPGA 1G platform using ReClick to verify their design and their router was able to process packets at the 1 Gbit/s line rate of NetFPGA 1G.

PacketC [Jun12] is a language developed by Cloudshield [Dun09] to be used on their heterogeneous multiprocessor machines for high-speed network packet processing (e.g., CS-2000 [Clo08]). These systems consist of FPGAs, processors and specific hardware (e.g., TCAMs). The programming model of PacketC uses coarse-grain, SPMD (single program, multiple data) parallelism to free users from thread management. On the hardware, FPGAs (e.g., for ingress filtering and header evaluation) and special microcode controls the packet pipeline. Single instances of a packet program are then executed on network processing cores. PacketC is based on C for the basic syntax (operators, conditional statements etc.), while difficult operations for a hardware implementation, such as pointers, address operators and dynamic memory allocation, have been removed. For packet processing, PacketC adds domain-specific data types and operators.

In contrast to hardware compilers, Chimpp [Rub10] follows a more general approach in that it relies on an XML description for the composition of arbitrary packet-handling hardware blocks. These blocks can be of various granularity (e.g., ARP lookup or simple TTL decrement), but they must be manually provided in a synthesizable HDL. Chimpp only supplies the interfacing / composition capabilities. The authors propose a basic library of modules with focus on routing applications for the NetFPGA platform, which they use to build an IP router and NAT gateway as sample application.

Gorilla [Lav12] is a C style language, that allows to describe data parallel applications (e.g., network processors) to be compiled on an FPGA on the functional level. Programmers can use domain-specific functions that are mapped to dedicated hardware accelerators. Gorilla defines these hardware accelerators as parameterized templates that have been written by hardware experts and optimized for high performance operation. From the top level, Gorilla follows a similar concept as Chimpp, but offers more fine-grained control of the application. The Gorilla compiler takes the application description and the templates and generates synthesizable Verilog code. The authors verify their approach on the example of a 100 Gbit/s Internet router, which can compete with up-to-date network processors in that domain.

Soviani et al. [Sov09] propose a solution for automatic synthesis of general packet processors based on packet editing graphs (PEG). A PEG is a compiler-style intermediate representation for the operations performed by a packet processing module. A PEG describes arithmetic and logical operations in a hardware-like style. Due to its complexity, it is intended that a PEG is generated from a higher-level language, however, Soviani et al. did not focus on that.

PEGs follow a dataflow-like structure with pipeline stages that each consume and produce a single packet stream. PEGs are automatically synthesized in hardware using a finite state machine, always trying to assure that within each cycle one word of data is processed. Note that analogue to [Kar10], for the implementation also Soviani's work split up header and payload upfront and do only process the header, which needs to be taken into account for comparison of performance results. For evaluation, the authors developed header-processing modules for a gigabit-capable passive optical network fiber-to-the-home system that achieve data rates of 25-40 Gbit/s on a Virtex 4 FPGA.

3.4.3 Micro-programmable and customized Hardware

As an alternative to the generation of dedicated hardware from high-level languages, another approach is to generate custom programmable hardware that is targeted to the application domain but allows to be quickly updated to match changing requirements.

Brebner and Kulkarni employ micro-coded data paths [Kul06] for implementing a NAT application on an FPGA. The idea behind the micro-coded data path is to implement processing elements that are stripped-down to the minimum required set of function blocks to reduce the control complexity. E.g., it may not be required to implement a full arithmetic logic unit (ALU) for a packet processor. However, the experiments of Brebner and Kulkarni show that the degree of programmability has a direct impact on the required resources. They admit that more complex examples might get worse performance than using a regular CPU.

NetThreads [Lab09], which also targets the NetFPGA platform, goes one step further and defines specialized 4-way multi-threaded processors on the FPGA, which are then software-programmable in a GPL. However, the performance of the system does not reach the performance of dedicated hardware accelerators for complex tasks. Here, the flexibility clearly has a negative impact on performance. E.g., for a sample application that does regular expression matching to classify HTTP packets, a performance of roughly 2000 Packets/s is given in [Lab09], which is comparable to a throughput of approx. 16 Mb/s for 1024 B packets. Without dedicated hardware accelerators for regular expression or protocol processing, it appears questionable that the approach has performance benefits above those of real network processors.

That such an approach can create powerful solutions for a sharply defined limited domain has been demonstrated by Brebner and Attig with their hybrid solution for programmable packet parsing [Att11]. It consists of a domain-specific programming language and a compiler that generates custom micro-programmable hardware and the corresponding microcode. The proposed packet parsing (PP) language provides a high-level way of describing fields of packet headers and rules how these fields should be processed.

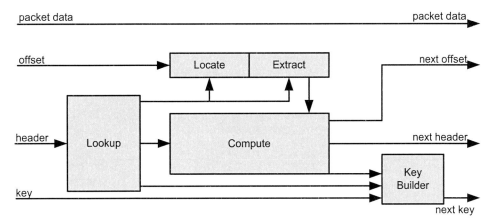

Figure 3.4: Micro-programmable hardware generated by PP compiler [Att11]

PP is based on the Java syntax for class definition, whereas a class defines header fields and methods for a particular protocol. PP descriptions are compiled into hardware using basic micro-programmable constructs for lookup (fetching microcode for particular packet), locate (find the interesting field), extract (extract data) and compute (calculate values). Figure 3.4 shows the connection of the different components. Brebner and Attig's generated example application uses 2048 Bit wide data paths to achieve a 400 Gbit/s parsing throughput on a recent Virtex 7 FPGA.

3.5 Honeypots

Honeypot systems can be grouped into two categories: low and high-interaction honeypots [Pro07]. While a high interaction honeypot provides a real system including a full operating system and applications to the attacker, a low interaction honeypot only emulates parts of an operating system (e.g., the network stack) and / or parts of a vulnerable application. Some literature also introduces a third category, the medium interaction honeypot, which contains more application logic than a low interaction honeypot, but not a full operating system. In the three-category scenario, the low interaction honeypot is further limited to the basic operations on the transport layer [Wic06].

High interaction honeypots help to learn much about an attacker's behavior when he is trying to get unauthorized access to a networked system and are often used in research environments, where continuous system monitoring can be provided. Low- and medium-interaction honeypots in contrast are well suited to assist a network IDS by providing information about worm activity, automated attacks for particular applications or spam attempts. Furthermore, they can be used to automatically collect malware that is trying to infect the honeypot using the offered vulnerable application emulations. Such honeypots are also called malware collection honeypots. Due to the relevance for this work, the following covers low- to medium-interaction honeypots only.

3.5.1 Honeypot Systems

Popular open source low- to medium interaction honeypot systems are HoneyD [Pro04b], Nepenthes [Bae06] and Dionaea [Dio11]. While HoneyD provides more a generic framework for the development of honeypot scripts, Nepenthes and Dionaea primarily focus on collecting malware using emulated application vulnerabilities.

HoneyD simulates a network stack that is able to forward packets to different service emulations. This network stack is furthermore configurable (using a personality engine) to change various fields in the header of packets, so that fingerprinting tools (e.g., Nmap [Lyo09]) assume that they are connected to a particular version of an operating system. HoneyD provides network routing capabilities, so that large IP address ranges can be assigned to HoneyD systems. Service emulations can be attached as simple applications and can be written using any programming language.

Nepenthes provides a platform that supports the implementation of honeypots with multiple emulations of vulnerable services. In contrast to HoneyD, Nepenthes focuses on the emulation part of the honeypot and uses the network stack of the operating system. The architecture of the Nepenthes platform consists of vulnerability modules (that emulate vulnerable applications and receive malicious requests), shell code parsing modules (that analyze the received malicious code), fetch modules (that can download malware from remote locations, if an URL is extracted from a malicious request), and logging modules (that monitor the operating of the honeypot). Nepenthes comes with emulations of popular vulnerable services, e.g., an IIS vulnerability emulation, and multiple worm backdoor emulations. The core and the emulations are written using C++.

Dionaea is the successor of Nepenthes and currently under active development. Dionaea follows the same goals as Nepenthes by collecting malware spreading around automatically. However, Dionaea especially focuses on emulating the

Microsoft SMB protocol, as it has a long history of bugs that are exploited by worms. In general, Dionaea tries to provide very detailed emulations of vulnerable applications so that various attack paths can be analyzed. In addition to SMB, Dionaea provides HTTP, FTP, MySQL, MSSQL and SIP emulations. Dionaea modules are written using Python code, whereas the core is written using C. Note that the SMB emulation of Dionaea is also used by mwcollectd [Wic10], another low interaction honeypot.

As an alternative to the various open-source honeypot tools, honeypot functionality is also offered by commercial applications. Specter [SPE02] provides a honeypot system that emulates a set of basic services. However, a special feature of Specter is that it contains functionality to get information about the remote attacker by probing the remote system itself. Furthermore, Specter can even leave invisible marks on the remote computer for later analysis by law enforcement agencies.

A further use case is shown by Mykonos. In their web security application suite [Myk12], a user is transparently directed to, e.g., a broken copy of the web application running in a sandbox, after he has been detected as acting malicious, so that he cannot harm the system any more.

HoneyBox [sG09] offered by secXtreme is a honeypot appliance that can be directly placed in a production network to gather information about malicious activity. It contains a pre-configured low-interaction software honeypot application running on the appliance hardware (which is a regular server) that emulates various network services. The advantage of the appliance is the easy integration into existing environments for small- and medium-sized companies, that do not have an own IT department.

3.5.2 Hardware Honeypots

To the best of the author's knowledge, there is only one published result about a concept to transfer a honeypot to bare hardware. The work of Pejovic et al. [Pej07] describes a hardware honeypot implementation by interpreting state machines stored in memory. These FSMs represent conditions and actions to describe the client-server interaction [Lei05]. The group implemented a hardware prototype using a Virtex-4 FPGA, but unfortunately did not publish any performance benchmarks.

Furthermore, their network stack is not fully complete and they talk about implementing parts of it on a PowerPC CPU. While the table-driven FSM approach is easier to program than a dedicated hardware approach, it might

become a bottleneck due to limited memory bandwidth when network speeds of
10+ Gbit/s are considered. Also, the flexibility of the solution is restricted to
the operations provided by the FSM engine. In that, the concept is comparable
to those of micro-programmable hardware. However, this work has not been
developed further.

4 NetStage Core Architecture

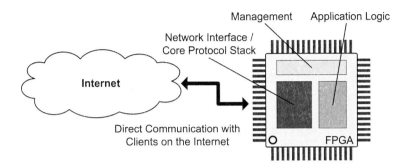

Figure 4.1: NetStage operation scenario

The implementation of active high-level network security applications entirely on dedicated hardware requires a base architecture that supports autonomous Internet communication on the FPGA without any further subsystem (e.g., a CPU). The basic principle of such an architecture is shown in Figure 4.1. A network communication core provides its service to custom application logic inside the FPGA, while a system management unit provides supporting control and management services.

As said, a major component of such a communication platform is an efficient implementation of the basic Internet protocol stack [Bra89], with particular focus on the transmission control protocol (TCP) [Pos81c]. Due to the lack of available solutions at the time of making a design decision, the development of a new implementation (NetStage) has been started [Müh10c, Müh10a, Müh10b].

NetStage is a cost-effective platform that supports both the User Datagram Protocol (UDP) and connection-oriented TCP processing at 10G line rates on FPGAs by using a lightweight design. NetStage supports millions of concurrent active connections even on target hardware with limited resources [Müh10c]. In addition to the network communication core, NetStage does offer additional resources that support current FPGA technologies often used in combination with FPGA network application, such as DPR for time-multiplexing of hardware resources.

With its integrated concept and its explicit design around a fully hardware-based Internet communication core, it distinguishes itself from other hardware-based networking research platforms currently available (see Section 3.2). These advantages make NetStage an ideal platform for research on FPGA-only high-speed applications for today's 10G networking environments.

This chapter presents the overall NetStage platform architecture, including the core architecture that has been presented in [Müh10a, Müh10b, Müh12b] and the lightweight TCP solution initially proposed in [Müh10c]. Details on application implementation, supporting services and platform scaling will follow in the next chapter.

4.1 Platform Design

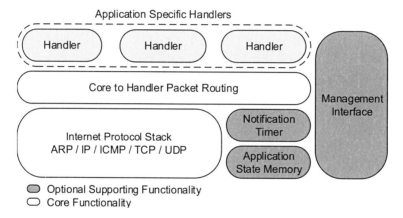

Figure 4.2: Components of the NetStage platform

NetStage implements the endpoint-oriented communication shown in Figure 4.1. A fully hardware-based TCP/UDP implementation, the NetStage communication core, provides its service to applications on the FPGA. Furthermore, additional supporting functionality assists the designer when implementing applications on the platforms.

NetStage consists of a modular platform architecture with multiple components (shown in Figure 4.2), that can be combined matching the current application demands. The components can be divided into two classes: core components and optional components. Core components are essential to establish communication sessions on the platform, while optional components add additional functionality.

The set of core components consists of the NetStage communication core as well as the packet routing layer that attaches custom applications to the core.

Additional modules for scheduling time-based events (Notification Timer), a global connection-based application-level state storage (Global Application State Memory) and an external management interface that allows monitoring and control of the system independently of the production network complement the platform as optional components. Furthermore, two scaling mechanisms allow the extension of the platform beyond a single FPGA: time-multiplexing of hardware resources by DPR and spanning the platform over multiple FPGAs. Details on this are given in Chapter 5.

In its basic form, all components have been developed to operate without any external memory while still providing high-speed communication scaling to millions of concurrent connections. This allows performing experiments using multiple different, inexpensive target development boards. Should more resources be available (e.g., fast external memory), additional functionality can be optionally enabled. Details on that will be given when describing the components.

Within the platform, the applications on the FPGA that use NetStage services are called Handlers. Fundamentally, Handlers are comparable to network endpoints / servers. They are independent hardware blocks that act as virtualized application managers. Single connections are processed on the hardware in an interleaved manner. The Handlers contain all application-specific functionality and are independent of other Handlers on the system. The number of concurrent online Handlers per system can range from one (for a very specific application) up to one hundred for large-scale multi-FPGA setups, depending on the available FPGA resources. NetStage does not explicitly provide the capability to forward data to a host CPU, since this is not in the scope of the current research. However, a Handler could also contain a Microblaze embedded processor [Xil11a], for example.

The Handlers are attached to the core using a unified interface (see Section 5.1.2). Developers can thus create application-specific Handlers without modifying the base architecture. The Handlers can be developed using, e.g., VHDL or Verilog code on the register transfer level (RTL) written by experienced hardware-developers. Another option is to use dedicated compilers that automatically generate RTL descriptions of Handlers out of higher-level languages. The advantage of such an approach is that the users of the system do not need to have hardware development experience. A prototype for such a compiler targeting the honeypot domain is given in Chapter 6.

NetStage has been implemented on two target reconfigurable computing platforms: The BeeCube BEE3 quad-FPGA prototyping platform and the NetFPGA 10G development board (see Section 2.2 for details). Both systems provide multiple 10G network interfaces and one of the larger Xilinx Virtex 5 FPGAs. While the NetStage core architecture itself is easily portable (assuming the target hardware has at minimum an FPGA with sufficient capacity and a network interface), these two platforms allow to focus on different implementation aspects. The stand-alone BEE3 has multiple FPGAs connected on a single board, and the NetFPGA card contains multiple high-speed external memories, SFP+ connectors for flexible network connectivity and comes as PCIe expansion card, that allows flexible housing. Initially, development was started on the BEE3 and later expanded to the NetFPGA 10G card when it became available.

4.2 Platform Architecture

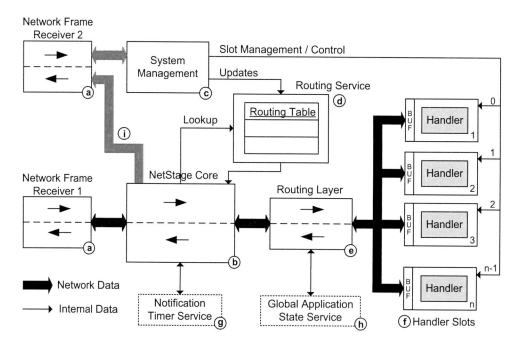

Figure 4.3: Core NetStage platform architecture
©2014 IEEE. Reprinted (modified) from [Müh14].

Figure 4.3 sketches the design of the NetStage platform architecture. The major building blocks are the communication core (Fig. 4.3-b), the routing layer

(Fig. 4.3-e), the application Handlers (Fig. 4.3-f) and the management interface (Fig. 4.3-c). Two dedicated network interfaces (Fig. 4.3-a) allow to separate network traffic between public Internet and internal management traffic.

Packets that arrive at the system from the external interfaces are processed by the core and forwarded to the corresponding application Handler. NetStage supports multiple applications on the same FPGA. Routing the packets to the right application after they have been processed by the core is the task of the routing layer (see Section 4.3.5). An internal routing table holds the destination information for each application. Destination lookup is performed by the routing service (Fig. 4.3-d).

When operated as network communication endpoint, the routing rules frequently correspond to network sockets, identified by the 3-tuple destination IP address, protocol and port. But the system is flexible, other routing schemes could be easily implemented (e.g., based on source addresses or even packet payload) by replacing the routing table with a custom implementation. The routing table can be dynamically updated by the management interface or simply statically set at compile time.

To increase the flexibility, the application Handlers are not directly attached to the core, but instead placed into Slots (see Section 4.2.3). These Slots provide the connectivity between the routing layer and the Handlers. As they encapsulate the Handler interface, they simplify the integration of new Handlers into the system, especially when partial reconfiguration is used. The operation of the Slots (e.g., enabling/disabling) is controlled by the management interface.

The connection between the Slots and the routing layer is implemented as a shared bus, supporting direct communication between the core and any Handler. This setup reflects the endpoint-oriented communication expected for the majority of cases allowing to simplify the routing layer implementation (see Section 4.2.5). The communication between the core and the different Slots is furthermore decoupled using buffers to avoid stalls in the main data path. For flexibility, these buffers are implemented as special ring buffers (see Section 4.2.2) in contrast to the FIFOs often found in flow-based networking platforms (e.g., [Nao08b]).

The core data path is clearly separated between the receive and send direction of packets to support full duplex operation at the native interface speed. As a special option, the communication core has a unidirectional connection to the management interface (Fig. 4.3-i) that allows sharing the network core between Handlers and system management components for sending IP-based control messages over the private management interface. This link could be of course dropped if security requirements prohibit such shared use of resources.

The internal data bus has a width of 128 Bit. Typically, the core is running at the same clock frequency of the network interface, which is 156.25 MHz for 10G [Xil11b]. Using a single clock avoids crossing clock domains and simplifies FPGA placement and routing, which has a positive impact on the design performance [Lam07]. The combination of width and clock rate leads to an internal data path throughput of 20 Gbit/s for each direction (incoming and outgoing), which is sufficient to satisfy a line rate of 10 Gbit/s while providing some headroom for throughput variations in the core.

The optional components Notification Timer (Fig. 4.3-g) and Global Application State Service (Fig. 4.3-h) are connected directly to the NetStage core and routing layer. Details will be discussed later in this chapter.

4.2.1 Message-based Communication

Even though NetStage uses traditional buses for internal communication, a message-based communication scheme is established to forward network packet data between modules. As common for such platforms ([Loc07]), NetStage messages encapsulate packet payloads by prefixing them with an Internal Control Header (ICH, see below). Such a message-based communication perfectly fits the packet-oriented communication performed by the hardware-accelerated applications and allows transmitting internal control data together with the packet payload in a seamless way.

Furthermore, the message-based interface is easily extensible: New core functionality can be added by simply changing fields in the ICH without the need to add extra signal lines to the hardware implementation. And any type of system-internal message can be transported over the same data channels as regular network traffic, thus saving hardware resources. The overhead of adding the ICH is to some extent balanced by the earlier removal of the packet protocol headers (see Section 4.3).

4.2.1.1 Internal Message Control Header

In contrast to providing only routing and control information, the idea behind the NetStage ICH (see Figure 4.4) is that the ICH should contain all necessary data for constructing a response packet without relying on any connection data stored within the core. Keeping the core completely stateless, saving both the storage as well as the access logic, allows an efficient data path implementation that offers high communication throughput at low resource requirements. While

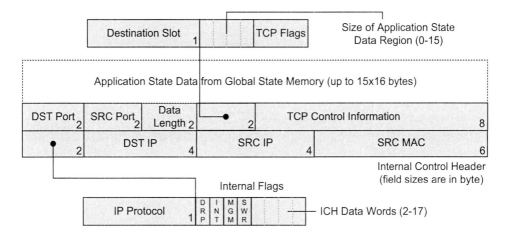

Figure 4.4: Internal Control Header (ICH) preceding each NetStage message

this scheme lacks support for some intricate protocol functionality (especially related to TCP, see Section 4.4), the prototype demonstrates that it is sufficient for experiments under various present-day network conditions.

The ICH generally encapsulates both incoming and outgoing traffic. In the current implementation, the basic ICH consists of two 128b words that contain packet source and destination addresses, optional TCP control information, and system control flags. Optionally, the ICH can be extended with a variable-size application state section to transport current per-session state data from the Global Application State Memory to the Handlers. Details on that are given in Section 4.2.6.

Table 4.1 lists the ICH fields and their meaning. Table 4.2 provides the explanation of the flags used in the ICH.

4.2.2 Buffer Design

NetStage message buffers are organized as ring-buffers, which support arbitrary addressing but avoid fragmentation. Figure 4.5 shows the buffer construction [Müh10a]. Messages are always stored contiguously in the buffer. An additional control information word contains the start address of each message in the buffer and its length in bytes. This allows the consumer and the producer of messages to read and write at arbitrary positions inside the packet by simply setting the corresponding pointer address.

Table 4.1: Internal control header fields

SRC MAC	The MAC address of the network device that forwarded the packet to the NetStage core.
SRC IP	The source IP of the client.
DST IP	The destination IP on the FPGA platform.
IP Protocol	The IP protocol for the current packet.
TCP Control	TCP control information (sequence numbers) used by the lightweight TCP implementation.
TCP Flags	TCP control information (flags) used by the lightweight TCP implementation.
Data Length	The raw data length of the NetStage message payload (excluding ICH and application state).
Dest. Slot	The destination Slot where the message should be routed to.
SRC Port	The source port of the client.
DST Port	The destination port on the FPGA platform.

Table 4.2: Internal control header flags

INT	Internal Flag: This is an internal message that should not be sent out on external interfaces.
MGM	Management Flag: This is a management packet to be sent out over the management interface.
SWR	State Write Flag: Data in the ICH application state region is valid and should be written back to state memory.
DRP	Drop Flag: This message has been marked as invalid (e.g., because of a wrong checksum) and should not be delivered to a Handler, but needs to be forwarded to maintain pipeline consistency.

This is, e.g., beneficial if a checksum needs to be calculated and added to the header after writing the packet to the buffer (avoiding local buffers in the modules) or if particular bytes somewhere inside the message need to be matched to decide whether it should be processed or discarded. If there is no match, the entire message can be instantly dropped by simply accessing the start address of the next message. If such a message would be located in a FIFO, it would need to be completely read even as it is not processed. While this is often not an issue for platforms processing network flows (e.g., all packets fly through the architecture),

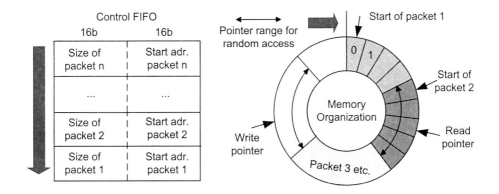

Figure 4.5: Design of the NetStage ringbuffer
©2010 IEEE. Reprinted from [Müh10a].

NetStage applications process only those particular packets destined for their service and all other packets arriving at the network interface should be filtered out as early as possible to conserve processing bandwidth.

The buffer has been implemented as follows [Müh10b]. A dual-port BlockRAM (BRAM) is used to store the message data. This allows producers and consumers of packet data to work independently, greatly improving the overall performance. A FIFO is used to manage the addressing and to notify the consumer of the availability of a new message. Each FIFO entry consists of a 16b start address and a 16b length field. The presence of an entry indicates that a complete packet is available in the BlockRAM. This assures that the consumer does not access locations of the message buffer that have not yet been written. Note that this approach (comparable to store-and-forward) does increase the latency of a single packet compared to directly starting to read after the first byte is available (similar to cut-through switching). The store-and-forward like approach does allow greater flexibility in the operation of the actual Handlers, as they can always assume a complete packet was received. With the research focus of this work, this was deemed more important than reducing latency.

In addition to the FIFO and the BlockRAM, each ring-buffer contains a buffer management unit (BMU) that keeps track of the number of bytes read and written to control the buffer fill status, This is important as the address pointer simply wraps around if the end of the BlockRAM is reached and would overwrite existing data if further data is written. When the number of bytes currently stored in the buffer exceeds the maximum capacity, a buffer full signal is asserted.

The buffer full signal is internally logically connected to the write enable signals of the BlockRAM and the FIFO in order to avoid data corruption due to partially overwritten data if a module does not correctly consider the buffer full signal.

The BMU offers an additional benefit that is especially important for an efficient use with multiple Handler Slots. Instead of requiring that the producer keeps track of the current write pointer for each connected buffer, the BMU holds the write pointer for the next free word in the buffer. Modules that write to the buffer simply use relative addressing when writing data and do not have to consider absolute addressing of data in the buffer. The BMU adds the correct offset to each write address, so that it maps to the free space in the buffer.

The size of the buffers can be flexibly set by sizing the amount of BlockRAM. The current maximum size is 65536 Bytes, limited by the 16b address and size fields of the control word in the FIFO, which should be sufficient for most application scenarios. The FIFOs are operated in first-word-fall-through mode [Xil12a], reducing the cycle overhead when reading new messages from the buffer.

Pipelining is used inside the core Handlers to further reduce the cycle overhead that is introduced by the two-step approach of FIFO and BRAM. As new messages typically start directly after the previous message, Handlers can prepare to read this data in advance while the previous packet is still being processed. Only if the FIFO control word indicates that this speculatively prefetched read was wrong, the read operation would need to be started again at the provided correct location.

4.2.3 Reconfigurable Handler Slots

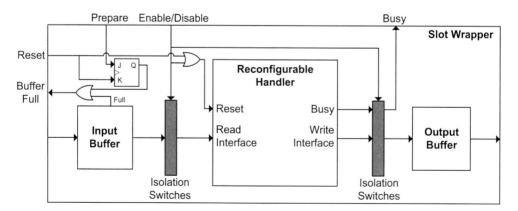

Figure 4.6: Slot wrapper architecture

Handler Slots encapsulate the raw application Handler [Müh10b]. The purpose of these Slots is to provide a single connection point and to enable FPGA DPR. As explained in Section 2.1.3, partial reconfiguration allows changing previously defined regions (partitions) of the FPGA during runtime without affecting the remaining FPGA logic.

All Slots implement the same Handler interface. As a system can have multiple Slots, a single Handler can therefore be allocated to any Slot if that Slot partition contains sufficient logic resources for that specific Handler. In that manner, dynamic partial reconfiguration (see Section 5.3) can be implemented to support time-multiplexing of hardware resources.

To simplify the integration of Handlers, a statically configured wrapper (see Fig. 4.6) encapsulates the Handler Slot. This wrapper contains the send and receive buffer logic for each Handler Slot and is not reconfigured along with the Handler. Furthermore, the wrapper contains enable/disable switches that isolate the entire Handler inputs and outputs from the system during reconfiguration to avoid unpredictable activity because of uncontrolled signal changes while the FPGA partition is reconfigured. As this isolation is not provided by the FPGA, it needs to be implemented explicitly in user logic.

The operation of the Handler Slots is controlled by the management interface. For this purpose, dedicated signal lines go from each Slot to the management module (see Table 4.3).

Table 4.3: Slot control signals

Enable/Disable	Enables or disables the connection between the buffers and the Handler. Furthermore, disabling a Handler Slot holds the attached Handler in the reset state, which initially resets newly deployed Handlers after they have been completely configured.
Prepare	Notify the Slot that it will be reconfigured in the near future so that it can perform cleanup tasks. In the current implementation, prepare sets a lock on the buffer that avoids any further messages to be written to the input buffer. In that fashion, the Handler can process the remaining messages in the buffer before being reconfigured.
Busy	This signal is high, if the Handler is currently processing a message and should not be reconfigured.

Note that the Slot Wrapper is also used if partial reconfiguration is not implemented, since it provides the input and output buffers. However, in that case the control signals can be tied to default values if they are not explicitly required.

4.2.4 Core-to-Handler Shared Bus

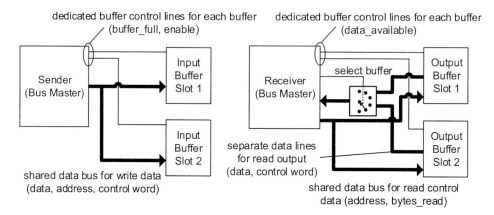

Figure 4.7: Architecture of the Handler communication bus

The routing layer connects the Handler Slots and the NetStage core. For efficiency, NetStage currently allows communication only between Handlers and the core, but not between individual Handlers. This limitation avoids the need for more flexible, but larger communication structures such as crossbar [Lee04b], mesh [MA07] or the adaptive CoNoChi [Pio08] as it is used for Dynacore [Alb10]. Furthermore, the core is the only master device on the bus. Therefore, a simple shared bus has been selected as the underlying communication infrastructure. However, NetStage runs the message-based communication on top of the shared bus, leading to an efficient solution for the Handler - Core communication. Such a tailored solution has also been chosen by [San09] to avoid complex routing schemes. The implemented bus allows unicast, multicast and broadcast of messages to the Handlers, whereas unicast is the preferred communication for active network communication Handlers. The other communication modes are useful for, e.g., passive Handler types (see Section 5.1.1).

Figure 4.7 shows the construction of the bus. The bus directly connects to the input and output buffer logic of the Slot wrappers. A bus sender and receiver module controls the operation of the bus. As in the core, the routing layer supports full duplex operation by providing two separate buses for receiving and sending packets.

On the incoming path all signals for the input buffers are shared among the Slots (the individual buffer-full signals being an exception). Therefore, data is present at all Slots during a write operation. The destination buffer is selected using dedicated enable signals for each buffer, controlled by the routing information contained in the ICH. In combination with the relative addressing scheme of the buffers (see Section 4.2.2) this setup allows easy implementation of unicast/multicast/broadcast to any combination of Slots.

On the outgoing path, the data signals of the output buffers are connected to a multiplexer that forwards the selected signal to the Slot aggregator. To select the appropriate source buffer, each data available signal is directly routed to the Slot aggregator which performs arbitration. Details on the implemented arbitration for NetStage are given in Section 4.3.5.

4.2.5 Routing Service

CORE ROUTING TABLE

Addr.	Slot	Handler ID	Pkt. Counter
(9b)	(8b)	(16b)	(16b)
1	3	4	1258
2	4	7	3659

MATCHING RULES TABLE

Addr.	State Words	Protocol	Rule Group	Port	IP	Netmask	Rule ID	Next Rule
(10b)	(4b)	(8b)	(16b)	(16b)	(32b)	(32b)	(16b)	(16b)
1	0	6	1	80	0.0.0.0	0.0.0.0	15	0xFF
2	1	6	4	23	1.2.3.4	255.255.255.255	4	256
256	1	17	2	23	17.5.2.0	255.255.255.0	52	257
257	1	6	24	23	1.2.3.0	255.255.254.0	12	0xFF
...	...	...	...	...	...	...	...	...

Figure 4.8: Structure of the routing table

The Routing Service manages the distribution of messages to a destination Slot holding an appropriate handler. Messages are routed based on their network endpoint information, but other schemes (e.g., including content-based routing) could be implemented.

For greater flexibility, the routing table [Müh11a] is split between a Matching Rules Table (MRT) and the Core Routing Table (CRT). The MRT (Figure 4.8-a) contains the endpoint socket identification as 3-tuple (destination IP address, protocol and port) and each incoming network packet is matched against this table. As additional feature, the MRT allows to specify the destination IP address together with a netmask, which enables the system to listen to entire ranges of IP addresses, a capability which is especially useful for, e.g., the honeypot scenario. Furthermore, the MRT contains the number of state words required by a particular Handler implementation. This configuration information is stored together with the routing information, since the value is required when packets are forwarded to the Slots to correctly attach global application state data to the message (see Section 4.2.6).

The CRT (Figure 4.8-b) contains the routing information, to which Handler a packet should be delivered. The combination of MRT and CRT supports multiple rules for a single Handler. The MRT contains a rule group ID as foreign key that links to the corresponding row in the CRT. The CRT further contains a packet counter field that counts packets arriving for that particular Handler.

For management purposes, the MRT contains a unique Rule ID that allows editing of particular rules with a management software. The next rule column of the MRT is used to speed up rule lookups. This process is described in more detail in the following Section.

The tables are stored in BRAM. In the current implementation, the MRT has a total width of 180b and the CRT has a total width of 64b. Unused bits are reserved for temporary experiments and future use.

4.2.5.1 Fast Hierarchical Routing Lookup

As the routing information is essential when forwarding packets to Handler Slots, the lookup process should provide a result within a minimum of cycles to achieve a throughput of 10 Gbit/s or more. For example, in order to achieve a 10 Gbit/s throughput, the lookup process for a 100 byte sized message should not take longer than 13 cycles (since 64 bit are received in each cycle on the external interface [Xil11b]).

While the routing rules table could grow too big (100 ...1000 rules, depending on the application) for a simple linear search, it remains too small for the complex large-scale lookup algorithms (e.g., as used in routers and firewalls with 10,000s

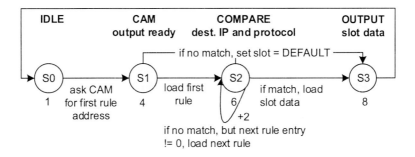

Figure 4.9: Look-up of destination Slot for a packet

of rules [Jia09, Nao08a]) to pay off here. Therefore NetStage implements a hierarchical rule matching algorithm (see Fig. 4.9) [Müh11a]. As network services are basically identified by their destination port, the port information will become the starting point for a chained rule lookup. All rules that contain the same destination port are grouped into a linked list. The next rule column of the MRT implements this list. The field contains the address of the next rule in the chain, or 0xFFFF, if the current rule is the last one.

If the first rule of a chain is known, the lookup process can easily go through the particular rules for a given port, which is much less effort than searching through the entire table. As the number of ports used in a system is expected to be at most around 200 different ports, a CAM is used to directly determine the starting address of the rule chain for a given port within a single clock cycle [Loc11] (additional cycles are added due to register transfer steps). The linked list further allows matching rules in an explicit priority ordering, considering more specific rules (having a longer netmask) before more general ones. For this purpose, rules will be inserted into the table depending on the length of their netmask. This is controlled by the rule management process. During lookup, the first rule that matches is chosen. After a matching rule has been found, the corresponding Slot information is fetched from the CRT by using the rule group ID. The Slot ID is then returned together with the state data control information.

The complexity of the rule lookup is therefore only dependent on the number of rules within a port chain, and not on the total number of rules in the table. It is expected that the number of rules for a single port should seldom exceed a dozen. E.g., for the honeypot multiple rules can be used to direct different network ranges to different or same target Handlers for the same application port. In the best case, a Slot lookup can be performed within 8 cycles (including wait cycles for memory fetches, see Fig. 4.9). Each additional rule in the linked list costs 2 additional cycles, until a match is found. If the CAM does not contain valid port information for a match request, the lookup can be quickly terminated with a cost of only 5 cycles.

For implementation efficiency, the head elements of each list are all located in the lower 256 entries of the MRT (to be directly addressable by an 8b CAM entry). As only the port numbers of existing rules are stored in the port lookup CAM, 256 entries suffice for the current implementation.

The design of the routing service (see Figure 4.10) easily supports the implementation of multiple parallel Slot lookup processes to further increase the throughput. For this purpose, the MRT, CRT and the CAM need to be duplicated, which could be easily done as the lookup process is read-only and all the tables can simply be written in parallel in case of updates. Finally, the results from the separate lookup processes need to be collected in a single queue. Due to the NetStage core pipeline implementation, it is important that the order of lookups is preserved. This could be achieved by not attempting to load-balance between the modules, but instead simply forward lookup requests on a round-robin basis and collect the results in the same order.

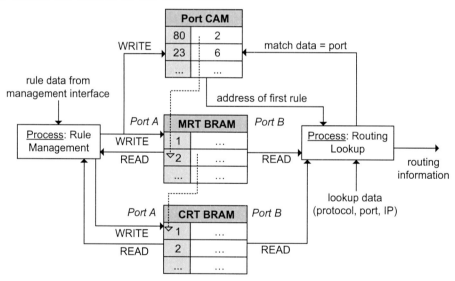

Figure 4.10: Implementation of the routing service

4.2.5.2 Rule Management

The Rule Management subsystem (Figure 4.10-b) accepts commands from the management network interface through a separate FIFO. Currently, the interface implements the insertion and deletion of rules. Updates can be implemented as a combination of delete and insert. An internal FIFO keeps track of available row addresses in the MRT. If a new rule should be inserted into the table, the rule management process first asks the CAM to determine whether a rule chain does already exist for this port. If this is the case, the process jumps into the linked list and goes through the entries until the right position according to the netmask information is found. The new entry is then inserted into the list by rewriting the next field of the predecessor entry.

If the port for a rule is not in the list, a currently unused address is allocated and the rule is written as new entry to the table. Furthermore, a CAM entry is made for that port pointing to the newly allocated address. The deletion of a rule is performed by rewriting the next address information of the previous entry to unlink it from the list. If the deleted rule is the only entry for that port, the CAM is cleared for that port by overwriting the CAM address. In the current implementation, the CAM can be shared between the two processes handling rule management and routing lookup, as rule updates occur only rarely and therefore do not affect routing throughput.

4.2.5.3 Pipelining

As packet forwarding will be delayed until the routing has been determined, the best case is that routing information is already available when the packet is ready to be forwarded to the Handler Slot. To avoid a large buffer that holds packets waiting for routing information, NetStage uses pipelining to perform destination lookup while the packet is processed by the communication core. For this purpose, relevant data for the lookup request (currently: destination protocol, port and IP) is extracted immediately from the packet headers when a packet enters the core and forwarded to the routing service as lookup request.

The routing service puts the result in a FIFO, where it can be picked up by the responsible core module. Note that this process strictly relies on retaining the same order between packets and routing lookups, so that the FIFO content matches the correct packet. The system guarantees this by not dropping a packet after a routing lookup has been issued, but instead using the drop flag to mark an invalid packet.

4.2.6 Global Application State Service

The Global Application State Memory (GASM) [Müh12b, Müh10a] has been introduced to maintain all application state data relevant for a single connection at a central place. This relieves the Handler from having to implement a memory interface itself, which would complicate Handler development and increase resource demands. Such a centralized storage is also more efficient than attempting to locally store state in each Handler (which would fragment the capacity of the on-chip memories). The global approach is very suitable to handle small volumes of state data that must only be maintained over a short time (e.g., session IDs or application context data). The application state data is transferred together with the network packet data by piggybacking it on the ICH. As this does increase the total packet size, the data is injected into the packet late in the processing pipeline, namely at the routing layer (see Section 4.3.5).

The Handler can then read the state information, process, and update it. The format of the state data is not fixed, each Handler can freely use the available data words to create its own record type. Any time the Handler sends out a message, it can request that the GASM be updated from the new values in the ICH Application Data Region by setting the SWR control flag. Note that always the *entire* global application state data region is written back to the memory, there is no way to mask individual bytes.

The access to the GASM is provided by the global application state service. This service encapsulates the read and write requests and allows to implement any backend for data storage (on-chip / off-chip). FIFOs on the input and output ports are used to buffer the requests / results. Two separate state machines independently perform the read and write requests. The state service currently accesses the QDR SRAM on the NetFPGA board using a memory controller provided by Xilinx. To achieve fast lookups, access to the data is implemented as direct mapped memory scheme, with data being identified using a hash value that points to a memory address. The hash value identifying a client session is generated using a universal hash function [Müh07] over the source and destination IP/port tuples:

$$H_k(M) = (m_1 \oplus k_1) \cdot (m_2 \oplus k_2)$$

with m_1 and m_2 being the lower and upper half of the message M, and k_1 and k_2 being the lower and upper half of a predefined key k.

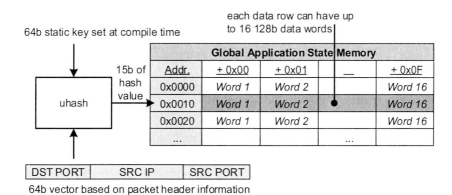

Figure 4.11: Implementation of the Global Application State Memory lookup

Figure 4.11 shows the design of the GASM and the lookup. In the current implementation, a cluster of 16 words (256 Bytes) is reserved for use by each session (limited by the available bits of the corresponding size entry in the ICH). Given the size of the QDRII-SRAM available on the NetFPGA card, this allows 32,768 addressable clusters (15b address vector).

Note that due to the reduction of the hash for accessing the GASM, collisions are possible and Handlers could get invalid state not destined for them. However, the live experiment demonstrated an equal distribution of hash values over time (see Section 7.4.1), which highly reduces the probability of a collision. It does not hinder our use-case of a hardware-accelerated honeypot, but other applications might need even more complex hash conflict-handling strategies. A further check could be implemented by storing the entire IP / port combination as part of the 128b word that is currently not usable by the Handler (as the ICH currently can transport at most 15 words of application state, while the GASM stores 16 words for each Handler). Whenever state data is retrieved, this value would be compared to the original full index and if the data does not match, an empty record would be returned. This approach would break the current session, but would avoid the delivery of incorrect data.

Furthermore, state data is never removed from the state memory, only over-written. Therefore, Handlers could inadvertently retrieve previous state data for a particular connection that is now invalid. To assure that Handlers start with a clean session state for a new connection, any established TCP session triggers a state write with empty session data.

While the global application state service has proven to be a solution for

multiple scenarios, designers need to consider the following idiosyncrasy: There
is a small gap (between a few and a hundred cycles, depending on the system's
buffer fill status), where newly arriving packets will be bundled with old session
data if the updated data has not yet been written back because the corresponding
message is still processed by the Handler or is waiting in the state write FIFO
queue. Figure 4.12 illustrates this issue.

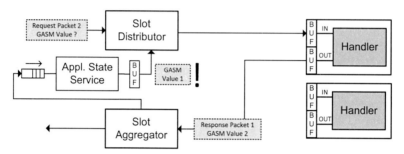

Figure 4.12: Potential delivery of outdated global application state data

While this interval of potential uncertainty is acceptable for the honeypot (as it
works conversation-oriented over the Internet, where the response arrives orders
of magnitude later than this delay), other applications might need to handle
it in a different way. An option could be to ensure proper synchronization by
implementing a local cache in each Handler buffer that contains not-yet-written-
back state data. If a new packet for the Handler arrives, a lookup would be done
in the handler-specific state cache, if updated state data is available. If yes, then
the data in the ICH would be replaced by the data in the cache. This is similar
to the store-to-load forwarding performed in conventional processor pipelines.

4.2.7 Notification Timer Service

In many cases, the approach of having the Handlers just react to incoming
packets is sufficient. However, there are scenarios where the Handler should react
independently of incoming external traffic, and instead wakes up after a set time
interval (e.g., throttling of connection speed for single clients). For this purpose,
the notification timer has been introduced [Müh12b]. The notification timer
simply stores notification messages (these are internal NetStage messages having
the INT flag set) and injects them into the receive data path after a given time
interval.

To provide space for notification messages, the implementation of the Notification Timer contains two FIFO queues implemented in one of the external SRAMs of the NetFPGA card, that are managed by two static timers that can be set at compile time. Working with a fixed timer avoids the need for rearranging or scanning the queues in their entirety [Nee10], which would require a significant number of resources to perform this task at acceptable speeds for hundreds of thousands of notification messages. The next packet is always the first element in the queue, together with an information about the time when it should be delivered. Figure 4.13 illustrates this concept. The use of a fixed number of timers is also common to other TCP hardware implementations [Int12b] to save hardware resources.

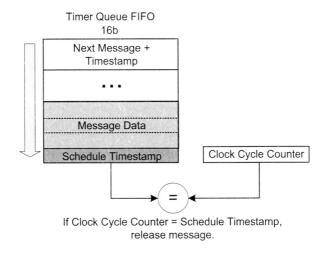

Figure 4.13: Notification queue design

A sample notification message could look like the one shown in Figure 4.14. It must contain at least a type classifier (denoting it as notification message) and the timer queue that should be selected. The ICH is implicitly included, so that a released notification packet follows the same routing principle as standard packets. This is crucial for operations such as DPR, which might have altered the contents of the Handler Slot that will actually receive the notification message sent by the earlier Slot contents.

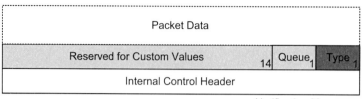

Notification Message

Figure 4.14: Structure of a notification message

4.3 NetStage Communication Core

The NetStage communication core is one of the base components of the NetStage platform. It handles the entire incoming and outgoing Internet data traffic and performs protocol processing up to Internet Layer 4 for the TCP and UDP transport protocols. NetStage implements ARP, IP, ICMP, UDP and TCP processing to support the basic set of communication protocols for implementing a wide range of applications.

In contrast to a software implementation (where optional functionality can be provided using libraries), on hardware even rarely used functionality always requires hardware resources on the FPGA. DPR can be used to adapt the system on demand, but currently the performance of the reconfiguration process is not high enough to embed this technology into a communication core for high-speed operation (see Section 7.2.4). Therefore, hardware implementations do need to trade off between functionality and resources. Often, rarely used functionality can be omitted without adverse effects (e.g., fragmentation [Int12b]).

Such a trade-off evaluation needs to be made for NetStage. The system is to be suitable as a research platform for autonomous FPGA-only Internet applications that supports experiments at speeds of 10 Gbit/s or more for a high number of concurrent connections, while the hardware demands stay on an acceptable level. It should support a multitude of clients as well as multiple applications simultaneously. Finally, the following criteria guide the trade-off decisions. The corresponding implementation details and design decision are described in the following Sections.

- The main data path should be able to handle network traffic at least at 10G line rate.

- Multi-client capacity is more important than single-user performance (matching the desired application domain).

- Multiple applications and multiple network endpoints (sockets) should be supported on the FPGA.

- The platform should provide its basic functionality on hardware with very limited external resources (e.g., memory).

- For security reasons, all modules that have access to the main datapath need to be build using dedicated hardware.

- Communication is initiated from external clients only, not by the FPGA platform.

4.3.1 Core Architecture

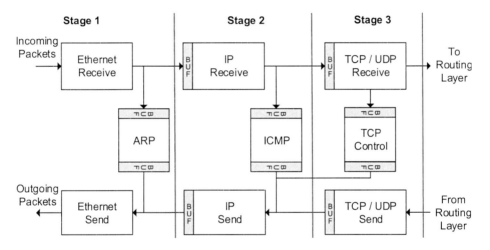

Figure 4.15: Architecture of the NetStage communication core

Figure 4.15 shows the building blocks of the NetStage communication core. Following the Stack of Internet protocols, the communication core has a modular and layered design, that allows to flexibly add new modules at various positions. This is a common principle for such network applications [Bra02]. Within NetStage, the different layers are called "Stages" which also inspired the name of the platform. The current suite of basic modules includes protocol-compatible implementations of ARP, IP, ICMP, UDP and TCP.

NetStage benefits from the large number of local memories available on today's FPGAs to implement buffers for the connection between all the core modules. This complements the modular design and decouples stages to avoid stalls in the main data path, if a module has temporary problems processing data at the highest speed (e.g., if a Slot lookup takes extremely long). As already said, separate data paths for the receive and send direction are used to enable full duplex operation at line speed. Directions are always seen from the core's point-of-view: receive means that the core receives packets from clients on the network and send means that the core sends packets to clients on the network.

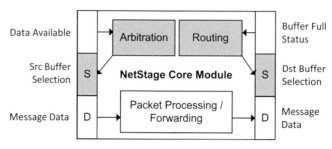

Figure 4.16: Design of a NetStage core protocol handling module

The modules for each stage are now discussed in greater detail. Basically, packets are forwarded between inter-stage buffers onward in each direction. For the buffers, the core uses the same ring-buffer like structure as the Handler Slots (see Section 4.2.2) to simplify the module development when assembling packets.

The basic layout of such a core processing module is shown in Figure 4.16. On the input side, the module reads packets from the inter-stage buffers. For buffer selection, each module can implement its own arbitration scheme. Currently, round-robin is used here for network traffic. On the output interface, processed packet data is forwarded to the next stage. As there can be multiple destination buffers, each module has local routing logic that is used to select the target buffer. Inside the modules, data is processed according to the stream-oriented principle often found in such packet processors [Loc07]. Data can be written to the output buffer while incoming data is still processed inside the module. This pipelining supports efficient forwarding of packets between the stages.

4.3.2 Stage 1 - Ethernet Layer

Stage 1 contains the Ethernet frame transceiver and receiver modules (see Fig. 4.17). Its external connections attach directly to the 10G network interface, implemented on both hardware platforms by using the 10G MAC [Xil11b] and XAUI [Xil10b] FPGA IP Cores from Xilinx. Stage 1 receives the raw Ethernet frame without preamble and CRC checksum, which has already been processed

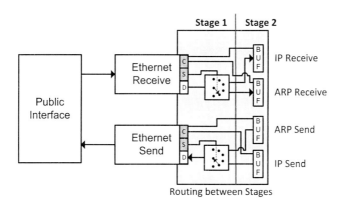

Figure 4.17: Details of NetStage processing in Stage 1

by the external IP cores. While the BEE3 version connects directly to the 10G IP core using the Xilinx LocalLink protocol, the NetFPGA version uses the newer AXI4 Streaming Interface to connect to the NetFPGA 10G interface wrapper provided by the NetFPGA 10G project. Attaching any other 10G interface can be done simply by adding another protocol interface converter at the external connection.

Packets that leave or enter Stage 1 on the internal connections do not carry the ICH to save processing cycles (as there are no protocol headers removed at this stage). Instead, Stage 1 always processes the entire raw frame (including the MAC address header). In contrast to the inner 128b data path the external interface has a data path width of 64b (defined by the external infrastructure) and the state machines that serve external connections are fully pipelined to operate continuously on the data stream without incurring any wait cycles. Thereby, a 10 Gbit/s throughput can be achieved.

Receive Path   The state machine in the Ethernet receiver module inspects incoming packets and forwards the packets without modification to either Stage 2 for IP processing, or the ARP Handler, based on the destination MAC address and type field in the Ethernet header. Packets are only accepted if the MAC address matches the configured local system MAC address (or additionally the broadcast MAC address for ARP packets), otherwise they are silently dropped. The local system MAC address is set during compile time.

Figure 4.18 shows how the destination address signals are generated inside the core modules. As the rules to select the next module are rather simple and fixed,

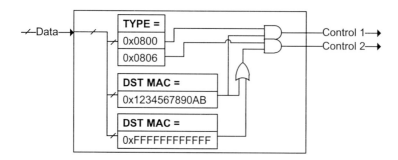

Figure 4.18: Routing classification of incoming packets in Stage 1

the logic is implemented statically instead of maintaining a dynamic routing table for inter-module packet routing inside the core.

Send Path   The transmit state machine implements the outgoing interface according to the chosen protocol standard (LocalLink / AXI4). It selects ready-to-send packets from one of the attached source buffers from Stage 2 and transmits them directly, only swapping the original source and the local MAC address.

4.3.3 Stage 2 - ARP and IP Layer

Stage 2 contains the IP processing modules and the ARP Handler (see Figure 4.19). This stage also manages the connection to the notification timer service (see Section 4.2.7). Notification messages are injected into the data path at this stage to trigger a Slot lookup, that is pipelined in the IP module.

4.3.3.1 ARP

Ethernet links use MAC addresses that identify a particular network interface or device for destination addressing. Therefore, the higher-layer IP addresses cannot be used for direct forwarding of packets on the physical layer. To deliver packets, devices need to know which physical destination is in charge of a particular IP address. This is handled by the address resolution protocol (ARP) [Plu82], that has been defined to link IP addresses with the physical interface address.

   A device that is intending to send an IP packet to another device attached to the same physical link transmits an ARP request message containing the destination IP to a specific physical layer broadcast MAC address. As a result,

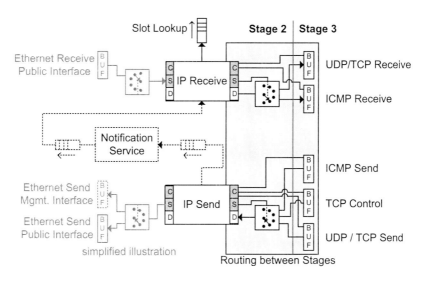

Figure 4.19: Details of NetStage processing in Stage 2

the device in charge of that specific IP address responds to this request with an ARP response containing its own MAC address.

As performing such an ARP lookup does slow-down the actual transmission of packets (as packets need to be held until the arrival of the ARP reply), network devices generally contain an ARP cache that holds the MAC address information for recently used IP addresses to directly send packets for known IP addresses. With its limited complexity, the ARP protocol could be easily implemented in hardware. However, the particular requirements of the honeypot application are better met by a more involved implementation that is described next.

Receiving Packets To avoid the need for static routes in the attached routers, NetStage implements a special *catchall* mode for high flexibility. The NetStage communication core responds with an ARP reply to *every* request for an IP address that comes in on the network interface. This avoids performing local lookups completely to check whether an IP address is in charge of the MalCoBox. Note that by responding to an ARP reply does not always mean that NetStage will actually process the packet. The selection of packets is based on the application routing table, which will be queried in later processing Stages.

In a production setup, the IP address range for the hardware honeypot can be defined by simply routing subnets to a particular router interface (where the honeypot is attached). If required, a lookup table could be implemented in the ARP module, that contains a list of IP / netmask entries that are explicitly *excluded* from the catchall mode. However, such functionality was not required in the experiments.

Sending Packets   On the transmit path, when NetStage sends response packets to clients, it also needs to know the MAC address of the next hop destination physical device. To avoid implementing a network routing lookup engine to select the next hop based on the destination IP, NetStage benefits from that it is currently designed to respond to incoming requests only. It stores the MACs of incoming packets in the ICH and uses this information when preparing response packets. This results in replies being sent back always to the physical device (e.g., a router) that originally forwarded the incoming packet, an approach that works well except for rare load-balancing/fail-over topologies using asymmetric routing. Same mechanisms are also used in commercial network equipment with hardware acceleration (e.g., Auto Last Hop by F5 [Net12c]) to speed up packet transmission.

4.3.3.2 IP

The Internet Protocol (IP) [Pos81b] is the base protocol for the majority of communication on the Internet. The main purpose of this protocol is to route packets between hosts over long-distance Internet links passing through multiple routing nodes. The protocol header contains the source and destination IP address (public IP addresses are worldwide unique numbers), that are used to identify sender and receiver. Furthermore, the IP header contains the protocol type of the payload and a checksum. Note that the checksum is only protecting the header itself, not the payload. Additional fields for specific transmission functionality complement the header (e.g., time to live (TTL) or type of service (TOS) for priority routing).

A special issue with IP packets is fragmentation. Every physical link has a certain limit of bytes (called maximum transfer unit, MTU) that it can transmit in one single packet. As an IP packet can travel over various different physical link types on its way through the Internet (broadband access, transatlantic lines, satellite links), there must be a way to deal with the situation where a packet comes from a link with a larger MTU an continues to a link with a shorter MTU. As packets larger than the MTU cannot be transmitted in one step on a link, such packets need to be split up into two or more parts. In the IP protocol, these parts are called fragments. The process to split up larger packets into smaller ones is defined in the IP standard and is called fragmentation.

However, fragmentation has a significant drawback. Before a fragmented packet can be processed at the destination, all fragments need to be available. This introduces requirements for large buffers, which is especially costly for a high-speed hardware implementation, where the processing pipeline should not be interrupted. Furthermore, in case that a single fragment is lost, the whole packet needs to be dropped. Furthermore, fragmented traffic induces security issues [Pta98, Zie95]. That is why fragmentation is avoided whenever possible in modern network environments (e.g., by MTU path discovery [Mog90]), because upper level protocols (such as TCP) can handle variable packet sizing (e.g., called segmentation for TCP) much more efficiently.

As the resource requirements are disproportional to the benefit, NetStage does not support fragmentation. Instead, NetStage relies on the reassembly functionality of the attached router, should fragmented packets arrive from the Internet.

Receiving Packets   Processing IP packets mainly consists of evaluating the IP header and filling the ICH with the source and destination IP address, the protocol and the source MAC. This information is mandatory when a response packet is created for this request and can be also used for logging purposes (e.g., to log an attack to the honeypot). Finally, the messages are forwarded to either the TCP/UDP module or the ICMP service based on the protocol header field.

In addition to the IP processing, the destination Slot lookup process already begun in this stage for any packet that matches the protocols implemented in Stage 3 (currently: TCP and UDP). This pipelining reduces buffer requirements, since the Slot lookup process is performed while the packet is still processed in the IP module.

Sending Packets   Sending IP packets simply consists of creating an appropriate IP header based on the data in the ICH. The IP header checksum is generated using dedicated hardware and pipelined with the data processing, so that the checksum data is available when the header should be written to the outgoing buffer of the module. Here, the module benefits from the ring buffer structure in that it can extend the payload already in the buffer by later adding the header with the checksum in front of the payload.

As the current implementation supports two network interfaces (one for external public and one for management traffic), the IP Send Module contains a separator that directs outgoing packets to either one of the buffers based on the MGT flag in the ICH.

4.3.3.3 ICMP

The Internet Control Message Protocol (ICMP) [Pos81a] has been designed to transport control information for Internet communication links between hosts. It is based on the IP protocol. The set of messages contains e.g. a notification, that a particular service on a host is not available or that a packet could not be forwarded because its size is too large for the MTU of a certain link and fragmentation is not allowed. However, for security reasons, nowadays most ICMP messages are prohibited by the firewalls of service operators.

One of the ICMP messages that is of particular interest to the NetStage platform is the ICMP Echo request (also known as Ping). A host that is configured to deal with ICMP Echo requests sends an ICMP Echo response back to the sender to signal that this host (more precisely, a particular IP address) is active. This is helpful for testing purposes and essential for the honeypot use case: Many attackers first scan an IP address range using Ping requests to find hosts to target. To this end, NetStage currently implements Ping as the only ICMP message type.

ICMP Echo responses are generated by copying the incoming request payload into the response and constructing a proper header. Furthermore, a checksum needs to be calculated, which is done using dedicated hardware. While the ICMP protocol might be a candidate to be also run by an embedded GPP, the clear security decision to implement all modules with direct connection to the main data path using dedicated hardware prohibits such an option here.

4.3.4 Stage 3 - UDP/TCP Layer

Stage 3 contains the UDP and TCP processing modules (see Fig. 4.20). In contrast to [Müh10c], the current version of NetStage has combined processing modules for incoming and outgoing UDP and TCP messages. As core operations can be shared between both protocols (reading from buffers, forwarding messages), this reduces hardware resources.

Before packets are processed by either the TCP/UDP Receive Module, the routing lookup result for that packet needs to be available, as the next output buffer is selected based on the state information (with or without state) and packet data should be immediately written to the output buffer when read from

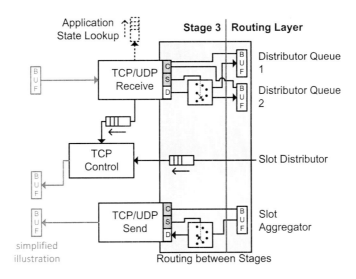

Figure 4.20: Details of NetStage processing in Stage 3

the input buffer. Furthermore, packets are dropped if there is no valid destination Slot, to avoid wasting processing power for useless packets.

4.3.4.1 UDP

UDP [Pos80] is a simple transport layer protocol on top of the IP protocol that can be used to exchange messages between services. The UDP header contains a source and destination port number that identifies services on a host. The UDP protocol itself is stateless and provides no transmission guarantee (applications can implement this on their own, if required).

UDP processing is done straight forward and is easily implementable in hardware [Ala10]. The UDP Receive Module just strips the header and extracts the source and destination ports, passing it into the ICH. Accordingly, the UDP Send Module constructs the UDP header based on the ICH data for outgoing packets. The UDP checksum is currently not implemented to save hardware resources, since this is optional according to RFC 768 [Pos80].

4.3.4.2 TCP

In contrast to UDP, TCP [Pos81c] is a stateful protocol which establishes a reliable communication channel between two communication partners on top of the (itself unreliable) IP protocol. Like UDP, TCP uses port numbers to distinguish between different applications behind a single IP address.

The reliability is based on two counters on each side, one for the cumulative number of bytes received and one for the cumulative number of bytes transmitted since establishment of connection. TCP implementations divide the data to be sent into TCP "segments". Whenever a TCP segment is transmitted, it carries the current byte offset of the data within this segment relative to the beginning of the connection as part of the TCP header. This number is called the sequence number (short SEQ). The receiver compares it with its own stored number of received bytes and can thus detect missing bytes, which have not yet been received (e.g., due to packet loss).

If all expected bytes were received, the receiver notifies the sender of the successful in-order reception of bytes by sending its current number of cumulative received bytes including the actual ones as acknowledgment (ACK) back to the sender. This is done by a packet without payload (called acknowledgment message) that has the ACK number as part of the TCP header and asserts the ACK flag. The sender can compare this to its number of bytes transmitted and verify that all outstanding bytes have been successfully received by the communication partner. If the sender does not receive the acknowledgment for bytes outstanding after a certain time interval, it will send them again (called retransmission), as it assumes that the original segment was lost and has never reached the receiver.

The drawback of the reliability is the need to track various state data on both the sender and the receiver. As this needs to be maintained for every single connection and processed for every packet received or sent by the system, processing a large number of simultaneous connections requires significant processing power solely for managing state information. Since this processing is especially costly on software systems [Mak04], common Denial of Service Attacks (DoS) aim to bring down systems by simply overwhelming them with TCP connection requests (SYN flooding). Due to the fact that TCP processing induces large performance costs on the server, research on ways to relieve the server from this work has been performed by various groups. The proposed solutions range from avoiding session state tracking [Hay10] up to new protocol proposals that direct work to the client, e.g., Tickles [Shi05]. The drawback of the latter is, that it is not compatible with standard TCP clients, which limits its usability.

Hardware implementations have been proposed that rely on external memory to store the connection state information [Sch02]. However, they require hardware platforms with fast external memory to store large numbers of concurrent connection state data items (> 1 Million). As an alternative, NetStage proposes a lightweight hardware implementation [Müh10c] similar to the concept of [Kam02] that uses the actual TCP header information to reconstruct the current session state. The major advantages are that this implementation is able to support millions of concurrent TCP connections at 10G+ rates without any external memory requirement. Note that, while such an implementation is compatible with the standard TCP implementation, there are some issues that arise from the absence of local session state memories. The stateless approach is discussed in more detail in Section 4.4.

The TCP implementation for NetStage consists of three modules: the send and receive modules, as well as a dedicated control module that handles TCP control messages (currently ACK and FIN messages). The TCP Control Module has a connection to later stages, as, e.g., it might be required to inhibit ACK messages if destination buffers are full (so that the packet is retransmitted by the sender). ICH flags that support the TCP implementation are listed in Table 4.4.

Table 4.4: Internal control header TCP flags

CLS	Close Flag: Close connection implicitly with this packet. Set by a Handler when preparing a response packet.
CES	Connection Established Flag: Signals a newly established connection. Important if, e.g., the client waits for an initial greeting message from the Handler directly after connection establishment.

Receiving Packets   Due to the stateless TCP approach, TCP processing does not require external data. The TCP process evaluates the type of the packet (SYN, ACK, FIN, DATA) and executes the corresponding actions. E.g., SYN packets trigger the 3-way handshake (described in greater detail in Section 4.4.1).

The interface between the TCP receive process and the TCP Control Module is a simple FIFO that contains the type of control information to be generated together with the ICH data to craft a proper response packet. Note that the entire 3-way handshake is performed without the application Handler being involved. Handlers are therefore totally freed from managing this task. However, Handlers are notified about the event of a successful handshake by an internal message with only the CES flag set.

If the packet is a regular DATA packet, it is forwarded to the routing layer. There are two destination buffers in the routing layer (see next Section): one for state and one for other packets. The destination buffer is selected based on the application state flag returned together with the Slot lookup result. This assures that packets that do not need state data are not delayed by the state lookup process. This results in a higher throughput for mixed traffic, if large portions of state are used for some connections.

Sending Packets The process for sending TCP packets consists of creating a proper TCP header based on the data available in the ICH (see also Section 4.2.1.1). Again, checksum calculation is supported by a hardware implementation that is pipelined with the data process.

4.3.5 Stage 4: Routing Layer

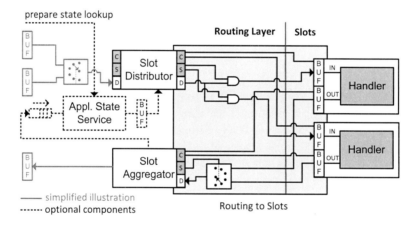

Figure 4.21: Details of NetStage processing in Stage 4

The modules in the routing layer (see Fig. 4.21) control the operation of the buses that are used to deliver messages to the Handler buffers and collect data for sending from them. Additionally, the routing module also inserts optional application state data from the GASM into the message. In contrast to [Müh10a], where only a small amount of state has been used (4b), for a flexible setup with more state data it is now added to the packet data as late as possible in the processing pipeline (to reduce forwarding overhead). Furthermore, fetches of global application state data should only be performed if required. As this information is stored within the routing table, that decision can be only made after Slot lookup (whose result is available in Stage 3).

4.3.5.1 Forwarding packets to Handlers

The message distributor receives messages to be forwarded to Handlers from its input buffers. When packets arrive at the routing layer, the ICH destination Slot field already contains the binary-encoded number of the destination Slot as an 8b value (see Section 4.2.1.1). If multicast is required, the upper half of the 256 addresses could be used to address previously defined multicast groups, either fixed at compile time or dynamically set. For systems with only a few Slots, the 8 bit value could also be interpreted as bit vector with each bit corresponding to a single Slot. This would allow any arbitrary combination of multicast groups.

The buffer full signals are evaluated by the Slot distributor before sending a packet to the Slot buffer. If the selected destination buffer is full, the packet will be dropped (which is a common principle for network systems). In case of TCP, after successfully forwarding a packet to the destination buffer, an ACK message is triggered in the TCP Control Module. To benefit from TCP retransmission in case of a full buffer, an ACK notification will only be sent to the client if the packet has finally been delivered to the Handler.

Note that in a multicast setup, a decision has to be made about what happens if a packet should be forwarded to multiple buffers and one of these buffers is full. This question cannot be answered for the general case and needs to be implemented in an application-specific manner.

4.3.5.2 Aggregating packets from Handlers

The message aggregator reads messages to be sent from the Handler output buffers. Buffer selection is done on a round-robin scheme for simplicity. Whenever a buffer has been selected, it gets a token and this buffer is only selected in the next turn if no other buffer signals that it has data available. The aggregation unit controls the multiplexer to select one of the attached Handler buffers.

The round-robin scheduler guarantees a maximum latency for a single Handler and assures that the bus is used to full capacity. All message transfer operations on the bus are atomic and cannot be interrupted. In this fashion, the round-robin scheme could lead to unfair situations, if one buffer has very large messages and another buffer has only very small ones. In that case, the buffer with the large messages would get a larger portion of the available bandwidth. If this proves to be a problem for particular application, other schemes could be integrated on demand (e.g., by including a priority [San09] or by using other algorithms, such as lottery [Sai11]).

Messages that are read from the buffers are directly forwarded to the network communication core attached to the routing layer. If a message contains application state data and the write state flag is set in the ICH, it is written back to the GASM by feeding this data together with an identifier (source and destination port and IP) into a FIFO queue at the global application state service (see Section 4.2.6). Furthermore, this portion of data is removed from the message before it continues downstream towards the Ethernet interface.

4.4 Lightweight TCP Implementation

As described in Section 4.3.4.2, the current NetStage TCP implementation reconstructs the servers SEQ number from incoming packets. This works especially well if all communication follows the request - response - next request pattern, which is, e.g., the case for the Honeypot. In the following Sections 4.4.1 - 4.4.5, details of this approach are described, partly repeating published material from [Müh10c].

4.4.1 Lightweight TCP Connection Establishment

Even in the lightweight variant, TCP connections are still established using the well-known "three-way handshake" [Müh10c]. Client and server exchange random initial sequence numbers (identified by a set SYN flag in the TCP header), and mutually acknowledge them. Random numbers are used to defend against connection hijacking attacks, where the attacker predicts the next sequence number and uses it to inject a packet of his own into the connection. Instead of storing the sequence numbers (which would again require memory accesses), NetStage uses a simplified mechanism (shown in Figure 4.22 [Müh10c]) similar to SYN cookies [Edd07].

The NetStage TCP implementation replaces the random number used as initial sequence number Y with a 32b hash value computed from source IP and port (called endpoint identifier, EID), and a secret key K. Assuming K remains secret and the hash function $H_k(EID)$ is strong, an attacker should not be able to hijack the connection by predicting the next sequence number. The system can check for the correct expected incoming sequence number $Y + 1$ by locally computing $Y = H_k(EID)$ again, instead of storing per-connection random numbers in slow external memory. This process uses the same hash function than for the GASM address calculation mentioned earlier (see Section 4.2.6).

In addition to the handshake, client and server can negotiate certain options during connection establishment. An important value is the Maximum Segment Size (MSS). This number defines the maximum number of bytes that will be transmitted as one IP packet. It can be up to 64 kB. However, practically used segment sizes are much smaller since the packet size of the underlying physical channel is limited. As fragmentation (see Section 4.3.3.2) could have a negative impact on TCP performance [Pos83]), sender and receiver negotiate a MSS, as the number of bytes (excluding protocol headers) that can be transmitted between them in a single packet without fragmentation. Due to this convention, in terms of size TCP segments are often identical to Ethernet packets (excluding headers) in current network environments. As the stateless implementation does not support dynamic MTU path discovery, NetStage assumes a fixed MSS of 1300 bytes as a safe value (based on results from [Mai09]).

Furthermore, advanced options such as window scaling and selective acknowledgment that rely on connection state data are not supported. Therefore, the only option currently implemented is the MSS (which in fact is also the only mandatory option according to RFC 793 [Pos81c]).

4.4.2 TCP Data Transmission

After successful connection establishment, data can be exchanged. To remain stateless, the outgoing SEQ and ACK numbers are computed from the incoming packet headers that accompany the packet in the internal control header (shown in Figure 4.22). While being compatible with the TCP protocol, this approach does have some limitations [Müh10c]:

- *Lost incoming single packets cannot be detected.* Here, NetStage can rely on the sender to just retransmit the packet after it has not been acknowledged by NetStage within the time-out period.

- *Packets arriving out-of-order or packets lost from a packet group cannot be detected.* NetStage attempts to avoid this situation by offering a constant receive window size equal to the MSS. Thus, at most one packet should be unacknowledged at a time and the sender must wait for the next acknowledge before it can send further packets. This is one of the cases where a deterioration of per-connection throughput is accepted to raise the number of concurrent connections. However, since recent results [Hur11] indicate that only around 5% of TCP segments are received out-of-order in current network environments, the overall effect of this issue should only be limited.

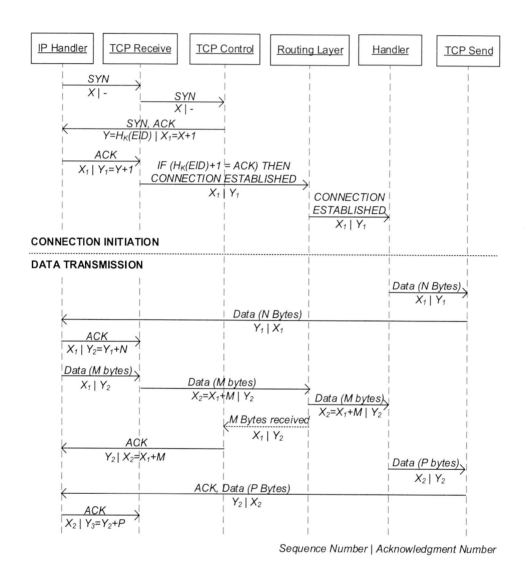

Figure 4.22: TCP sequence of the stateless implementation

- *Duplicate retransmitted packets are not detected.* This can occur, e.g., if the acknowledgment of a packet is lost. The packet is then just acknowledged again after it has been retransmitted, satisfying the sender. However, as the duplicate packet cannot be detected, it will be passed upward for processing a second time, possibly triggering actions in the Handler twice. Handlers need to be aware of this possibility. For the honeypot, this is easily manageable, as some Handlers are stateless or follow strict state machines that ignore packets arriving for the wrong state. Accidentally generated duplicate response packets will be discarded at the receiver due to SEQ and ACK numbers being equal to an already received packet.

- *Security* While supporting TCP communication with multiple clients on the Internet, the stateless approach does not cover the full security functionality offered by a more complete implementation. Attackers can simply send TCP packets with arbitrary SYN and ACK numbers, that will be forwarded to the Handlers, as the information that a successful connection establishment has been performed previously is not stored somewhere. However, this is currently not an issue for the research platform (as the hardware is not that vulnerable and no production services will run on the honeypot that could be affected by such attacks).

Furthermore, TCP retransmission, a functionality that assures reliable transmission of packets in case of packet loss, is not supported in the implementation. As sending retransmissions requires the information whether a certain packet has reached the destination within a certain timeframe, this could not be implemented without local state data. This is further justified by the observation that only about 10% of the TCP connections transferring less than 50 kB of data on the Internet suffer from packet loss at all [Mai09]. The experiments presented in Section 7.4.1 also validate this. However, retransmission functionality is easily added if sufficient external memory is available.

4.4.3 Window Sizing

To speed-up data transmission, implementations of the TCP protocol offer receive buffers allowing senders to transmit more than one packet without waiting for ACKs. For this to work, the sender needs to know the amount of free space in the receive buffer to avoid overflows. This is managed by the TCP header field "Window Size", set on every packet transmitted (data packets as well as acknowledgment messages), which indicates the number of bytes the receiver is ready to receive. If there is no buffer space left, the sender has to stop transmissions until it receives a packet indicating a new window size. NetStage currently ignores this, as it is assumed that for the honeypot the responses are rather small and that the clients should have enough time to process the packets.

4.4.4 Segmentation

The NetStage TCP core does not perform data segmentation when packets are processed in the TCP Send Module to keep inter-stage buffers small. This task is offloaded to the Handlers, which are themselves responsible for proper segmentation of larger response messages. This means that each Handler should split large portions of data into multiple NetStage messages that will be handled and transmitted as single packets.

Also when receiving segments, the Handler is responsible for managing data that crosses packet boundaries. If that is the case, intermediate processing results could be stored using the global application state memory, so that they are available with the reception of the next segment.

4.4.5 Closing Connections

Some Handlers may require shutting down the TCP connection after the last data packet has been transmitted. This can be signaled using the CLS flag in the ICH. The TCP Send Module then sets the appropriate flag in the TCP header.

The other case is that the remote station wishes to close down a connection by initiating the required handshake. Note that regularly connections always need to be terminated on both sides to free connection state resources. While this is not required for NetStage, since it simply does not store the connection state, NetStage imitates the handshake to free resources on the client by correctly shutting down the server side of the connection.

4.4.6 Comparison with other network communication cores

As stated previously, hardware TCP implementation do need to make some trade-offs to efficiently achieve high-performance results in comparison to software implementations. Table 4.5 gives a comparison of the features for three available commercial 10 Gbit/s TCP implementations and the NetStage lightweight design: DINI TCP Offload Engine [Gro12a], which is a single connection implementation designed for low latency, and PLDA QuickTCP [PLD12] and Intilop TCP Offload Engine [Int12b], both generic multi-connection implementations.

Table 4.5: Comparison of NetStage and other TCP cores

	Intilop	PLDA	DINI	NetStage
Implementation	full hardware	full hardware	hardware / CPU	full hardware
Protocols in HW	ARP, ICMP (Ping), IP, TCP	ARP, ICMP (Ping), IP, TCP, UDP	TCP data only	ARP, ICMP (Ping), IP, TCP, UDP
Concurrent Sessions	1024 (4000 optional)	16	1	unlimited
Listening Sockets	n/a	1	1	512 routing rules
User Interface	20Gbit/s (128b@156.25MHz)	20Gbit/s (128b@156.25MHz)	10Gbit/s (64b@156.25MHz)	20Gbit/s (128b@156.25MHz)
Operating Mode	Server and Client	Server and Client	Server and Client	Server
Max. supported MTU	1536B (9000B optional)	9000B	1536B	1536B
Hardware Timers	per Session (opt.) or group-based (default)	supported	supported	notification timer (group-based)
Reassembly	optional	supported	supported	not supported
Application Support	PCIe bridge	PCIe bridge	PCIe bridge	GASM, DPR

The implemented protocol set is largely identical, with the exception of the DINI core that handles only TCP data transfer in hardware and everything else using a CPU (even TCP connection establishment). Providing 20 Gbit/s of throughput at the user interface is also a common selection for the cores. The PLDA core is highly configurable and supports many features as optional selection, depending on the FPGA capabilities. In that, it is similar to the NetStage approach.

Major differentiation criteria are the number of concurrent sessions and listening sockets. These options have a significant impact on the core design (as state storage and lookup is required), so that while promising a major speed improvement amongst software implementations, on hardware the number of simultaneous connections is far more limited compared to even the capabilities of a commodity server. NetStage tries to lift these limitations and has been designed closer to the behavior of a real server in that regard. In turn NetStage is missing some functionality (e.g., reassembly), which would require huge buffers for the high numbers of connections the design is optimized for.

Furthermore, NetStage provides some additional application support functionality, such as the Global Application State Memory and the custom routing table, which might be difficult to implement on the other systems without source code access. In addition to providing internal connectivity for on-chip applications, the other cores do offer an optional connection to a PCI Express interface, so that they can be connected to a CPU for application processing. NetStage does not provide this now, as it has not been in the focus for the current work.

4.5 Chapter Summary

This chapter presented the core components of the NetStage platform. The NetStage platform has been designed to provide a development environment for hardware-based networking applications with the focus on active communication, which goes beyond the packet-level platforms that are already available. An implementation of the basic Internet communication protocols IP, TCP and UDP forms the central component that allows autonomous Internet communication without any software-programmable CPU involved. A flexible, message-based bus interconnect scheme allows to attach custom hardware modules at various stages. Fast routing of packets to destination endpoints inside the platform is supported by a hierarchical forwarding rule matching infrastructure.

The implementation of the TCP protocol follows a special lightweight approach.

To achieve a small footprint and to allow millions of simultaneous connections without requiring a large amount of costly high-speed memory, the TCP implementation uses a special stateless server mode. Connection state information is reconstructed from the header information transmitted with every packet. While this is very efficient to allow experiments on a wide range of affordable hardware, it lacks some functionality, e.g., retransmission. However, experiments show that this does not have a significant negative impact on the practical usability of the developed applications (see also Section 7.4), so that the benefits dominate.

5 NetStage Platform and Application Support

This chapter focuses on the platform aspects and the application layer of the NetStage platform. This includes details on application Handler development (Section 5.1), platform components and tools that support application development (Section 5.2), and dynamic partial reconfiguration (Section 5.3) as one way of scaling beyond the capacity of a single FPGA for additional application logic. An initial discussion of platform components has previously also been presented in [Müh12b] and DPR in [Müh11a, Müh12c].

The application Handlers can contain any arbitrary functionality. The only requirement is, that they connect to the message-based interface of the communication core presented in Section 4.2.1. To lower the development efforts for new handlers, a basic Handler template has been developed that will be used as skeleton for all application Handlers. This template contains, e.g., ready-made logic to properly read and write from the Slot buffers. Additionally, a special simulation environment that allows to directly connect the HDL simulator to live network traffic, eases practical development work. In that fashion, debugging can be performed using real interactive data, allowing the verification of complex communication patterns with multiple request / response parts.

The number of Handlers that fit on an FPGA highly depends on the application that is implemented. E.g., for the MalCoBox honeypot it could be estimated that around six average-sized Handlers fit on a current medium-sized FPGA (Virtex 5 TX 240T), before FPGA placement & routing will become an issue (see Section 7.2). While this will be sufficient for many scenarios, use-cases that require an even larger number of handlers are conceivable (e.g., large-scale honeypots with many vulnerability emulations).

An easy way of scaling the system is to use a larger FPGA device. However, at some point chip fabrication technology limits the size of the device and large devices often have costs that increase super-linearly with the logic capacity. Furthermore, the number of Slots is not only limited by the logic capacity of the FPGA, but also on the routing resources, as all Slots are connected to a single shared bus. To alleviate both issues, NetStage employs two approaches. The first one, DPR, exploits the capability of FPGAs to partially reconfigure the device with new functionality while the rest of the FPGA keeps operating (see Section 2.1.3).

DPR is a technology to *time-multiplex* some of the actual hardware resources that has been used in research projects to adapt the FPGA hardware to changing conditions [Cla10, He12]. To benefit from time-multiplexing, NetStage uses DPR to implement a hardware virtualization layer that can swap Handlers in and out based on the observed network traffic characteristics. In that fashion, the system adapts itself to the current workload.

As an alternative to the DPR solution, a second scaling option relies on the use of *multiple* FPGAs [Tan10, Son11]. Inspired by the ring-based design of the BEE3 quad-FPGA prototyping platform [Dav09], NetStage application Handlers can be distributed across multiple nodes, which are then connected in a ring structure, while only a single instance of the communication core is required in the entire system. Thereby, the available number of logic resources is increased.

The modular design of NetStage supports such an extension. In contrast to the DPR solution, all Handlers can be active at the *same* time here. This is beneficial for systems with compute intensive Handlers, but comes at the cost of a significantly more complex (and expensive) hardware platform compared to a single-FPGA system using DPR. For this reason DPR will be the more appropriate choice in most NetStage use-cases. This is reflected in this text, which omits the details of the multi-device architecture. Please see [Müh11d] for a thorough description of the quad-FPGA architecture.

5.1 Application-Specific Service Handlers

The service Handlers are responsible for the actual application-level processing of network data. The focus of this work lays on actively communicating services, where the application logic represents a communication endpoint. Fundamentally, the operation model of the Handlers is similar to a network server implemented in software. Such Handlers receive network packets on a specific socket and process the data received. In the following text, this type of Handler is also called Active Endpoint Handler (AEH) to differentiate it from Handlers operating in other modes, which will be discussed in Section 5.1.1.

Figure 5.1 shows the architecture of an example AEH for the honeypot service [Müh12b]. Note that in the honeypot application, AEHs are also called Vulnerabilty Emulation Handler (VEH) to better describe their concrete role in that use-case. Such VEHs react to incoming packets and generate response packets according to predefined rules. Their architecture consists of the actual protocol state machine, an (optional) regular expression matching engine, and an (optional) set of response packets described as stored templates. These three components need to customized for each application. Due to the regular structure, such a Handler for the honeypot application is also well-suited to be automatically generated by a high-level compiler, which will be addressed in Chapter 6.

Designers that develop Handlers for the NetStage platform are free to use any logic inside the Handler modules. They only need to ensure that the Handlers follow the basic operation principles of the NetStage architecture. These major constraints are:

- The interface to the communication core is fixed. All Handlers need to follow the message-based approach presented in Section 4.2.1.

- The current implementation of the TCP protocol targets specific application scenarios. The trade-off decisions given in Section 4.4 need to be considered.

- Handlers should avoid local storage of session data and use the global application state service instead, especially when DPR is used.

- NetStage currently only supports reacting to incoming packets (or notification events). Connections cannot be initiated by NetStage itself.

While the logic inside the Handler is application-specific, in most cases the Handlers will implement the stream-based processing that is also used for the communication core modules (see Section 4.3). When response packets are created, the original internal control header (see Section 4.2.1.1) needs to be remained.

Furthermore, to support the desired multi-threading to achieve high-performance, Handlers should be immediately available after a packet has been processed. Packets of different sessions, but using the same protocol, will be processed on the same hardware in an interleaved manner. Therefore, a single connection should never block the Handler.

5.1.1 Handler Types

While the focus of this work is on the AEHs, the architecture is sufficiently flexible to be used for other purposes. In summary, three different types of Handlers can be defined:

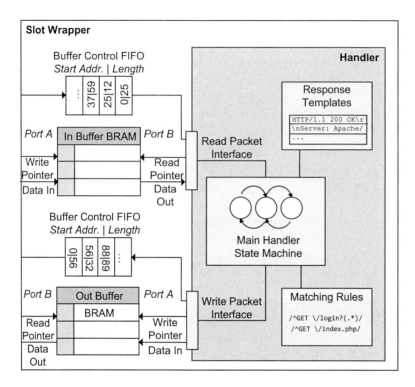

Figure 5.1: Example application Handler structure
©2012 ACM. Reprinted from [Müh12b].

- **Active Endpoint Handlers** (AEH) actively communicate with clients on the Internet.

- **Passive Endpoint Handlers** (PEH) passively monitor network packets and do not actively participate in a communication [Müh12c].

- **Active Inline Handlers** (AIH) actively inspect and alter packets that pass through the architecture on their way from an external sender to an external receiver.

The base functionality implemented in each Handler is identical for all Handler types. They receive network packets from the core, process data and send packets using the message based interface provided by NetStage. While the basic operation model is the same, the actual packet handling differs between the three types.

AEHs will send response packets back to the client, while the PEH only analyzes data (e.g., to track statistics) and will not send back any response packets to the original client. However, even a PEH can use its outgoing network interface to transmit status packets to a computer acting as management console. Even without performing conversations, PEHs benefit from the lower-level protocol processing to easily access header information, such as, IP addresses. Furthermore, the multicast and broadcast capability of the routing layer bus supports the use of multiple PEHs to monitor the same packets.

The situation is different for AIHs. Handlers of this mode will be placed inline in a communication channel between an external client and an external server. The current NetStage implementation uses only a single public network interface, thus packets will enter and leave NetStage on the same physical interface. Colloquially speaking, in this mode NetStage operates as a 'bump in the wire'.

The network could be configured to let AIHs process only packets that are going in one direction (e.g., the packets from the client to the server), while the return path from server to client is not monitored. The Handler could then process the data, for example by inspecting or changing payload data (e.g., encryption), and forward the packet to the real server by altering the destination MAC address in the outgoing packet (if the server is located in the same physical or virtual Ethernet network). Such a setup could be used, for example, to implement a custom hardware load balancer that distributes incoming requests to multiple servers (identified by destination MAC addresses) in a fashion similar to the Linux kernel loadbalancer IPVS [Pro04a].

Note that the NetStage ARP service does not need be modified for this mode of operation. The destination MAC address is carried in the ICH and can thus be also written by the Handler. For more complex AIHs, however, additional changes to the ARP and routing services as well as the routing table might be required.

5.1.2 Handler Interface

The Handler interface connects the Handler logic with the Slot wrapper (see Section 4.2.3). The interface consists of the following signals:

The main purpose is to read and write packet messages from and to the buffers. To support high-speed operation, each direction has a dedicated set of signals, so that packet data can be read and written in the same clock cycle. This is beneficial if data should be copied from the input packet to the output packet. The description of the signals is given in Tables 5.1 and 5.1. The timing diagrams

Listing 5.1: Handler Interface

```
entity PR_Handler is
  port (
    sys_clk                     : in   std_logic;
    reset                       : in   std_logic;

    -- receive data
    packet_data_in              : in std_logic_vector(127 downto 0);
    packet_adr_rd               : out std_logic_vector(15 downto 0);
    ctrl_fifo_rd                : out std_logic;
    ctrl_fifo_in                : in std_logic_vector(31 downto 0);
    data_available              : in std_logic;
    bytes_read                  : out std_logic_vector(15 downto 0);
    read_finished               : out std_logic;

    -- send data
    packet_data_out             : out std_logic_vector(127 downto 0);
    packet_adr_wr               : out std_logic_vector(15 downto 0);
    packet_wr_en                : out std_logic;
    write_finished              : out std_logic;
    ctrl_fifo_out               : out std_logic_vector(31 downto 0);
    send_buffer_full            : in std_logic;

    inactive                    : out std_logic
  );
end PR_Handler;
```

in Figure 5.2 and 5.3 show how data is read and written from the buffers.

An available packet in the incoming buffer is notified using the **data_available** signal. The Handler can then take the control FIFO information and start accessing the buffer memory locations holding the packet. The Handler should set the **inactive** signal low, if it is currently processing a packet to assure that it is not reconfigured while a transaction is in operation. Furthermore, Handler developers have to ensure that the **read_finished** signal is set properly to avoid orphaned bytes in the buffer after a packet has been read. If this is not considered, the buffer control logic does not free the memory leading to a deadlock situation quickly. The **bytes_read** signal needs to be set to the total number of bytes read from the buffer, including the packet payload and the ICH. In general, this is equal to the length value of the FIFO read control word **ctrl_fifo_in**.

The delay that exists between reading the FIFO word and then addressing the memory by using the start address from the FIFO control word (see Figure

Table 5.1: Handler receive interface signals

Receive Interface	
`packet_data_in`	128b data word read from buffer BRAM
`packet_adr_rd`	read address of next word from buffer
`ctrl_fifo_in`	32b control word from buffer FIFO that contains start address and length (in bytes) of next packet (see Section 4.2.2)
`ctrl_fifo_rd`	active high, fetch next control word from FIFO
`read_finished`	active high, acknowledge the message read and free the space in the buffer (set with bytes_read)
`bytes_read`	the number of bytes read from buffer (corresponds to the length value of the control word)
`data_available`	active high, if a valid buffer control word is available in the FIFO (which corresponds to a packet message)

Table 5.2: Handler send interface signals

Send Interface	
`packet_data_out`	128b data word to be written to buffer BRAM
`packet_adr_wr`	address of buffer word to be written (relative to start of message)
`packet_wr_en`	acive high, write enable for the buffer BRAM
`ctrl_fifo_out`	32b control word to be written when a message has been completely written to the send buffer. Contains start address (zero if relative addressing is used) and length (in bytes) of message (see Section 4.2.2)
`write_finished`	active high, acknowledges the control FIFO word and thereby notifies the next stage about the existence of a message in the buffer
`send_buffer_full`	active high if the send buffer is already full

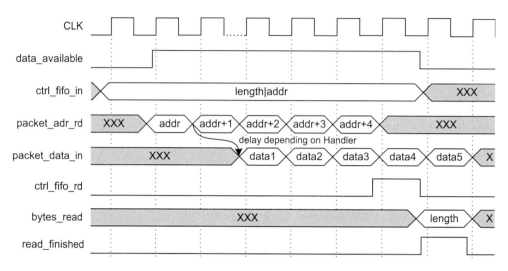

Figure 5.2: Handler signal timing when receiving a message

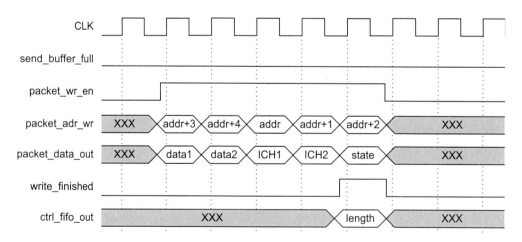

Figure 5.3: Handler signal timing when sending a message

5.2), which depends on the number of register levels inside the handler, can be easily compensated by using pipelining after the first packet message has been processed. As regularly the next packet starts at the last address plus one, the read address generation logic continues to request packet bytes and does only need to later verify if the start address is the same as it has speculatively expected and if not, restart.

Writing packets to the output buffer is straight forward. Handlers can use arbitrary addressing when writing data words. E.g., the developed Handlers write the packet payload first and afterwards add the ICH upfront currently. After finishing the packet, the Handler needs to write also the FIFO control word, containing the data length of the entire message including the ICH. `write_finished` denotes a valid control word and implicitly signals that the write of the message has been finished with this cycle. Writing the FIFO control word can be done together with the last data word to save cycles.

5.2 Supporting Platform Services

In addition to the optional components GASM and Notification Timer, NetStage provides a statistics counter and further supporting services that aid developing applications on NetStage.

5.2.1 Management Interface

The management interface provides central system control and monitoring functionality. It is currently implemented using dedicated hardware; however, depending on the complexity of the management functions, alternatively an embedded GPP could be used here. Since this part of the platform does not have a connection to the data path containing public (and potentially malicious) network data, this would not have a negative impact on the base security provided by the hardware-implemented network communication core.

Access to the management interface is currently provided by a dedicated network interface (up to 10G, if available on the platform) or by physically sharing the public interface using different MAC addresses for providing virtual interfaces (this should be used only for testing, as it physically mixes public and private network traffic). For simplicity, the functions implemented for system control and monitoring use raw Ethernet frames to communicate with the administration station (using a custom protocol version). Alternatively, a PCI Express interface could be implemented instead of using the network port.

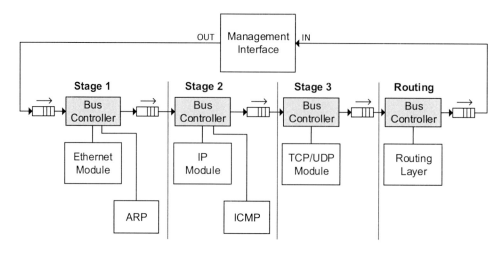

Figure 5.4: Implementation of NetStage management bus

The current set of control functions relates to managing the packet routing table (see Section 4.2.5) and Slot control (see Section 4.2.3). Rules can be added and removed by sending properly crafted packets. Furthermore, Slots can be manually enabled or disabled. The management interface can also be configured to send the current state of the packet forwarding table periodically to the administration station for display. For this type of notification messages, UDP messages are used. UDP has been selected to be able to easily implement listener applications on the management station. Please note that this breaks the operating principle that NetStage does not initiate connections by itself. As NetStage does not have an ARP resolver, the MAC address of the receiving administration station needs to be set manually in that case (see Section 4.3.3.1).

For the management bus, NetStage adopts the FIFO-based control message routing architecture from NetFPGA [Nao08b]. The control bus forms a ring, that starts and ends at the management interface and all modules requiring to process control messages are included in this ring. They contain local engines that process messages if they match the destination address (see Figure 5.4). Responses are then sent back using the same ring. While this is not efficient for high-speed communication, such a chain is resource-efficient. The management bus has a data width of 32 bit.

5.2.2 Extended Statistics

To obtain a better insight into the system behaviour under operation, an extended statistics option has been developed that can be enabled on demand. Extended statistics are collected for each core Handler and contains the following information:

- total number of messages processed since last reset (32b),

- total number of message bytes processed since last reset (32b),

- total number of packets dropped due to buffer overflow since last reset (32b)

- time spend on actually processing message data measured in cycles since last reset (64b).

The extended statistics are held in registers. Currently, each Handler has up to sixteen addresses for such statistics registers. Furthermore, additional measurements can be added as required. Please note that larger values (64b data) occupy two consecutive addresses. Furthermore, statistics data is only zeroed on global reset (e.g., during power on).

Statistics values are periodically queried by the management module using the management bus. During each data collection cycle, all addresses of relevant statistic data are read sequentially, with the addresses set at compile time in the corresponding VHDL process. Statistics values are transmitted as a single UDP packet to the management station.

5.2.3 Mirror Port Functionality

The NetFPGA implementation of NetStage contains an additional option for using one of the 10G interfaces as mirroring port for the public network interface. This supports debugging and monitoring of all network traffic if the system is connected to a public Internet uplink and no other port mirroring option (e.g., through the uplink switch) is available.

Incoming and outgoing packets on the public interface are simply duplicated and written to two FIFOs. The content of these FIFOs is then read and transmitted on one of the other network interfaces (e.g., the monitoring link) in an interleaved manner. As this is done directly in hardware, it does not have any impact on the system throughput. By connecting a server to the mirroring port, data can be easily captured and analyzed using standard system tools (e.g., tcpdump [tcp10]).

Note that because incoming and outgoing traffic is multiplexed onto a single link, the mirroring port is not able to deliver 10G of incoming and outgoing traffic. If this is absolutely required, another of the available four network interfaces could be used as additional mirroring port, with the incoming and outgoing packets directed to two separate interfaces.

5.2.4 Simulation Environment

In common practice, VHDL code is verified using pre-written test benches that execute a particular test case and record and/or check the response of the simulated system under test. This simulation is done using special simulation software, such as ModelSim [Gra10].

One drawback of this batch-oriented approach is the lack of interactivity. This is especially an issue for a platform such as NetStage, where interactive communication with external entities is one of the key elements. Fortunately, the simulation tool vendors do offer support for extensions that can be used to implement custom functionality.

Such an extension has been programmed for NetStage that supports interactive communication between a network interface and the simulator running the NetStage platform. This allows developers to verify the Handlers and inspect their internal behavior using actual communication partners (e.g., client software). The facility significantly improves the efficiency of developing with the platform, as existing programs can be used to test the system and changes in the HDL code can be quickly verified with the final communication partners. While this simulation does not achieve the high speed of the hardware (see Section 2.1.2), the functional verification significantly simplifies complex debugging on the hardware.

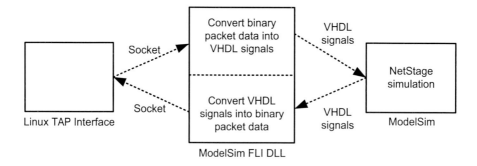

Figure 5.5: Attaching a virtual network interface to NetStage VHDL simulation

This interactive simulation environment has been developed using the foreign language interface (FLI) support from ModelSim, following a design example given in [Tig03]. Figure 5.5 shows the software architecture. The network interface has been implemented as virtual network interface on a Linux server (using TAP). The connection between the interface and the ModelSim simulator is established by a FLI program that converts binary packet data received on the network interface into VHDL signals according to the protocol expected by the NetStage core. The other direction is implemented vice-versa. In this fashion, the simulation setup skips the Xilinx XAUI and MAC core (for performance reasons, as simulating also these cores would be more resource intensive) and directly talks to NetStage, as the original MAC core would do.

5.3 Dynamic Partial Reconfiguration

Dynamic Partial Reconfiguration (DPR) implementations employ the on-chip reconfiguration interface of the FPGA to reprogram the logic resources (see Section 2.1.3). The process can be controlled either by an embedded GPP or by dedicated hardware, depending on the speed that should be achieved [Liu09]. Beyond the controller implementation, the type of bitstream storage affects the overall performance (e.g., Flash, SDRAM). Furthermore, the adaption algorithm controlling (scheduling) the reconfiguration activity is an important factor for the efficiency of a DPR implementation. Such algorithms can range from basic controller designs for static scenarios [Gar09] to highly specialized scheduling algorithms executed on an embedded GPP [Bau09].

NetStage employs DPR to support the dynamic replacement of Handlers [Müh11a]. The implementation allows holding number of Handlers in a repository that can be automatically deployed to the FPGA on demand. While this scheme can be extended to support different Handler types [Müh12c], the main NetStage version concentrates on active communication and has DPR implemented for AEHs.

On the technical level, the FPGA is partitioned into a number of reconfigurable rectangular regions that contain a balanced mix of processing elements (logic and memory blocks). These regions are reserved for holding the Handlers. A region thus corresponds to a Handler Slot (see Section 4.2.3). Figure 5.6 illustrates the basic principle of this hardware virtualization approach. Note that NetStage assumes that a single region is only used by a single Handler. Proposals exist for dynamic reconfigurable systems that build a grid of reconfigurable cells on top of the FPGA fabric and allow the placement of an application module on one or more of these grid locations [Koc08, Alb10]. While this allows arbitrary placement of differently sized modules, it induces issues such as fragmentation [Ahm10] and increased scheduling complexity: Now the scheduler not only has to manage the assignment of modules to fixed regions, but also to determine the optimal placement of these modules [Mah11]. Since the compiled NetStage Handler modules (see Section 7.2.2) are relatively close in size, the benefits of such an approach in relation to the increased complexity do not pay off here.

The following Sections 5.3.1 - 5.3.4 describe the implementation details, partly repeating published material from [Müh11a, Müh11c].

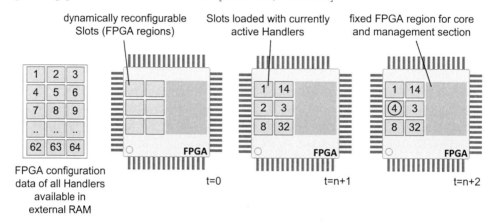

Figure 5.6: Hardware virtualization on NetStage using DPR

5.3.1 Architecture

Figure 5.7 shows the architecture of a system that includes the DPR service [Müh11a], that has now been merged with the latest revision of the NetStage core system. In addition to the module for NetStage DPR scheduling named Reconfiguration Candidate Selection (RCS, see Fig. 5.7-a) and the reconfiguration controller, named DPR Controller (DPRC, Fig. 5.7-b), the routing service needs to be updated to reflect the dynamic Slot changes initiated by the DPR scheduling service. Furthermore, statistics data that is required to perform autonomous adaptation decisions (see Section 5.3.3) needs to be collected by the communication core. The partial hardware configuration data files (partial bitstreams) are stored outside of the FPGA (e.g., in an external memory) and are loaded into the device by using the on-chip reconfiguration interface (ICAP) [Müh11a].

As the Handler Slots have been designed with DPR in mind and are already able to provide proper isolation of Slots during reconfiguration, no further changes are required here. On the functional level, the ability to dynamically replace Handler modules on NetStage imposes the following design rules:

1. A Handler should only be active while it is processing a packet, and should not be blocked by a single connection.

2. Session and state information for a connection (e.g., client transactions) is stored *outside* the Handlers in the Global Application State Memory provided by the static core architecture.

3. Execution of a Handler is only triggered by incoming packets, not autonomously (bypassing the scheduler).

Property 1 is essential to allow frequent swapping: A Slot is active only for a short amount of time (a packet is processed in a couple of microseconds). Property 2 ensures that no state has to be preserved when swapping Handlers. Finally, Property 3 guarantees that no packets are lost if Handlers are reconfigured, because they cannot wakeup themselves when they are offline. As both platforms (BEE3 and NetFPGA 10G) provide external memory and a Virtex 5 FPGA that has ICAP capability, the described approach can be implemented for both target systems.

5.3.2 Reconfiguration Controller

NetStage does perform the actual DPR operation autonomously of a host PC. Also, the DPRC performs the reconfiguration instead of an embedded GPP, since the latter would not be able to achieve high reconfiguration speeds due to its restricted memory bandwidth (see [Liu09]). The partial Handler bitstreams will be fed directly into the ICAP (see Figure 5.7-c) using fast DMA transfers. The implementation utilizes the current maximum ICAP speed of the Virtex 5 FPGA of 32Bit at a clock rate of 100MHz [Xil12b]. Experiments showed, that the ICAP interface can be overclocked to achieve even higher reconfiguration rates [Duh11, Han11]. However, as this is beyond the specification of the FPGA, it has not been adopted for NetStage.

With the used revision of the provided FPGA development tools (13.3), each Handler-Slot combination requires a separate bitstream [Xil11e]. This increases the storage requirements, as the total number of bitstreams that need to be stored

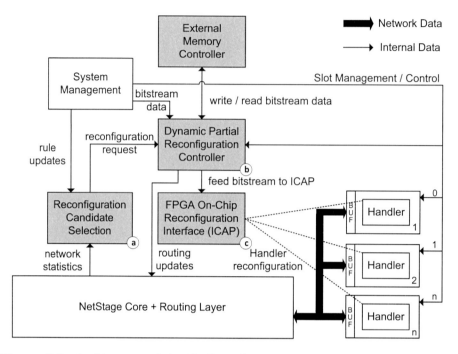

Figure 5.7: Architecture of the NetStage/DPR self-adaptable network platform

in the external memory increases exponentially if the number of Handlers and the number of Slots is increased. Even if the resources inside a reconfigurable FPGA region were completely identical, it is not possible to simply migrate a single Handler bitstream between different regions on the FPGA. This issue has been addressed by multiple research projects to support *runtime relocation* of bitstreams [Soh11, Cor09] in order to reduce the memory requirements by rewriting the addresses inside the bitstream to allow placement at multiple locations. However, while applicable for smaller modules, these solutions ignore timing issues that can arise from simply relocating modules without performing timing analysis [Bec12]. In 2012, Beckhoff et al. introduced their tool Go Ahead [Bec12], that provides an integrated solution to work with relocatable partial bitstreams and which particularly addresses timing. But the authors state, that in case of timing violations, even with Go Ahead more than one partial bitstream is required at runtime.

Furthermore, the restrictions regarding the routing of signals passing through relocatable modules, could cause a negative impact on the overall routability of the design. For maximum compatibility, the current NetStage implementation thus requires separate bitstreams for each Handler, corresponding to the different physical locations of the partially reconfigurable areas on the FPGA. The memory is organized in clusters, addressed by the Handler ID and the target Slot number (see Figure 5.8) [Müh11a]. This allows a simple addressing scheme where the memory location for a bitstream can be generated only from the Handler ID and the target Slot number. The configuration of the clusters depends on the number of Slots and the availability of memory. Furthermore, handler Slots are typically set to the same size to support arbitrary allocation of Handlers to Slots. Should external memory become an issue, bitstream compression techniques, such as runlength coding presented in [Hau99, Duh11], can be used to reduce the memory requirements.

The DPRC is only responsible for selecting the specific bitstream to load, the actual decision when and which Slot to reconfigure with which Handler is not made by the DPRC, but by the reconfiguration candidate selection (see below). The DPRC only receives the Handler ID and the target Slot number together with a reconfiguration request and then performs the reconfiguration in the following fashion:

1. Wait until the Handler is free by monitoring the busy signal.

2. Isolate the Handler by disabling the Slot switches.

3. Reconfigure the Handler by feeding (DMA-ing) the configuration data from the external memory into the ICAP.

4. Enable Slot switches after reconfiguration (releasing the Handler from its initial reset) and remove the input buffer lock (see Section 4.2.3) set by the reconfiguration candidate selection.

Note that the routing table is updated inside the RCS module, where the input buffer lock is set to avoid misrouted packets in the time between reconfiguration decision and actual reconfiguration.

5.3.3 Reconfiguration Candidate Selection

Reconfiguration Candidate Selection (RCS) is the process that decides which of the currently online modules (if any) will be replaced by a new module in each reconfiguration step. To achieve a high system throughput, the goal is to perform

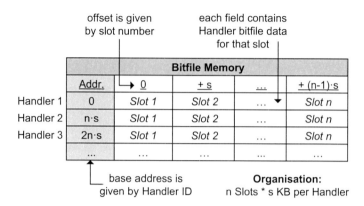

Figure 5.8: Addressing of Handler bitstreams in external memory

reconfiguration in a way that the majority of incoming packets (comparable to tasks in traditional scheduling) are destined for an already online Handler, so that costly reconfiguration activity is reduced to a minimum. Optimal solutions for this scheduling problem are still under research [Ste04, Bau09, Bau12, Alb10]. Replacement strategies can range from simple random selection over basic LRU (least recently used) schemes up to more refined algorithms that utilize additional data to make decisions.

Generally, the better the expected behaviour of an application is known, the better such an algorithm can be adjusted. As the prototype application for NetStage is the honeypot, the currently implemented RCS algorithm [Müh11c] is optimized to this application domain. Such a honeypot can have 1 to 100 different VEHs that emulate particular vulnerable services. The goal is to serve the maximal number of incoming requests with a proper VEH. As it is not possible to individually reconfigure a new Handler for each separate incoming packet at 10 Gbit/s (see Section 7.2.4.1), the algorithm should keep Handlers with a high rate of incoming packets on-line (configured in a Slot).

On the other side, the implementation of the algorithm should be kept resource efficient and fast, so that reconfiguration decisions can be made quickly enough to keep the reconfiguration controller busy. For this purpose, an algorithm based on the least-frequently-used (LFU) scheme has been selected that interprets packet statistics. It can be easily implemented on hardware. LFU is an algorithm that uses historical data to make decisions [Ste04]. For the statistics, packets are considered at the socket level (destination protocol, port, and IP address), accumulated over a set time interval (currently 1s).

Network statistics are collected by counting the number of packets that match the routing rules for a particular Handler. The algorithm (see Listing 1) scans this data set periodically to determine the off-chip Handler with the highest packet count, and the on-chip Handler with the lowest packet count. A second step checks if the high-count off-chip Handler has a higher packet count than the low-count on-chip one. If that is true, the high-count off-chip Handler replaces the low-count on-chip Handler (see Figure 5.6 at time steps t=1 and t=2). The network statistics are cleared periodically to remove Handlers that show no activity in the past period. This is done to avoid that Handlers block Slots just because a short but intense burst of packets boosted their counter to a large value in a previous statistics collection interval.

Using this algorithm, Handlers are kept on-chip that have the highest activity index in the past period. For the honeypot traffic characteristics, it is likely that these Handlers will also receive traffic in the near future. As Handlers are not blocked when participating in an active connection, they can be swapped in and out *multiple* times during an active communication.

A drawback of this algorithm is, that attacks which send only very infrequent requests might not be detected using this algorithm, as the required Handler will be swapped out before it has completely seen the attack, or is never even swapped in due to too infrequent traffic. The first weakness could be alleviated somewhat by simply giving each Handler a time-to-live (TTL) value that ensures it is present a minimum interval of time (to actually catch more of low and slow attacks). The second weakness could be addressed by having a higher level of decision-making based on longer-term statistics: It could ensure that Handlers for rarely occurring traffic do get a bump up in priority once in a while to make sure that every kind of traffic is at least occasionally observed.

5.3.4 Adaptation Engine Implementation

To build the statistics table holding the counters for the RCS, matching rules for each available (either on-line or off-line) Handler need to be stored. As these matching rules are similar in nature to the routing rules used for packet routing (see Section 4.2.5), both will be merged into the existing Matching Rules Table (MRT) which now holds the forwarding rules to *on-line* Handlers as a subset of the general matching rules (encompassing both on-line and off-line Handlers) [Müh11a]. To track the packet counts, the counter field of the Core Routing Table (CRT) is used to collect statistics per rule group. Off-line, but available Handlers are now indicated in the CRT by a target Slot value of zero.

Algorithm 1 Self-reconfiguration algorithm

---

deployedId, undeployedMax, undeployedId, i = 0
deployedMin = MAX

// search for replacement candidates
while i < CRT:numrows) **do**
  if (CRT:value(i) < deployedMin) **and** (CRT:slot(i) > 0) **then**
    deployedMin = CRT:value(i)
    deployedId = i
  else if (CRT:value(i) > undeployedMax) **and** (CRT:slot(i) = 0) **then**
    undeployedMax = CRT:value(i)
    undeployedId = i
  end if
  i++
end while

// trigger partial reconfiguration
if undeployedMax > 0 **then**
  if empty slots available **then**
    configure off-chip Handler(undeployedId) into free slot
  else if undeployedMax > deployedMin) **then**
    swap on-chip Handler(deployedId) with off-chip Handler(undeployedId)
  end if
end if

As they rely on the same data structures, the adaptation engine is realized together with the routing service in a common hardware module (Figure 5.9). It contains the logic for tracking statistics, interpreting rules, and managing Handler-Slot assignments. Due to the use of dual-port BlockRAMs, routing lookups can be performed in parallel to the counter process, as the routing lookup is read-only. Only the CAM is duplicated, so that the rule lookup and the counter lookup processes do not interfere. This increases the overall performance of both the counter and the lookup process.

The Slot lookup process as described in Section 4.2.5 remains unchanged. Note that as not all incoming packets should be counted (e.g., TCP ACKs should be ignored to count only real data packets), the adaptation engine uses a separate port (Figure 5.9-a) to receive counter update requests only for specific packets. The counter requests are generated inside the TCP/UDP Module of the core. Counter requests are currently generated for the following packet types / events: UDP, TCP Data and TCP Connection Established.

To perform reconfiguration candidate selection, the CRT is scanned periodically every 1 ms in the current implementation (this interval may not be shorter than the Handler reconfiguration time, see Section 7.2.4). After scanning the table, the values of the offline Handler with the highest packet count and the online Handler with the lowest packet count are immediately available. The decision reached will be used to update the CRT and to inform the DPRC (Section 5.3.2) to perform the actual reconfiguration operation. Reconfiguration requests are passed to the DPRC using the DPR Request FIFO as Handler ID and target Slot pairs. The delay that occurs between table update and actual reconfiguration is controlled via the prepare lock-in switch of the Slot wrapper. When the RCS has made a selection, the Slot input buffer of the Handler that will be reconfigured is set into the prepare state by using the prepare signal.

5.3.5 Discussion of Packet Routing

The current implementation of NetStage defines that packets having no valid destination will be dropped. Therefore, all packets that do not have a Handler online when the lookup process is performed, will be counted, but not forwarded and instead dropped. This corresponds to the simple but hardware efficient scheduling approach that tasks for which a reconfiguration unit is not available are rejected [Ste04]. A schedule incorporating future planning does not to be calculated in that case.

For TCP applications, this will be compensated by TCP retransmissions, so that it is likely that a Handler is online when the retransmitted packet arrives at the system. On the other hand, there is no guarantee that a packet reaches a Handler. E.g., for UDP packets there is no retransmission capability and therefore the packet might get lost forever. However, the common practice that attackers or bots trying to infect computers on the Internet often scan an entire IP range for open systems is beneficial for the honeypot here. This means that the honeypot can bring up an appropriate Handler after the first few requests for such a flooding attack and then have the Handler available for the requests arriving later. As Handler reconfiguration currently takes less than 1 ms (see 7.2.4), there is a high chance that some of the attacks will reach an online vulnerability emulation.

While dropping packets is common to networked systems if a target service

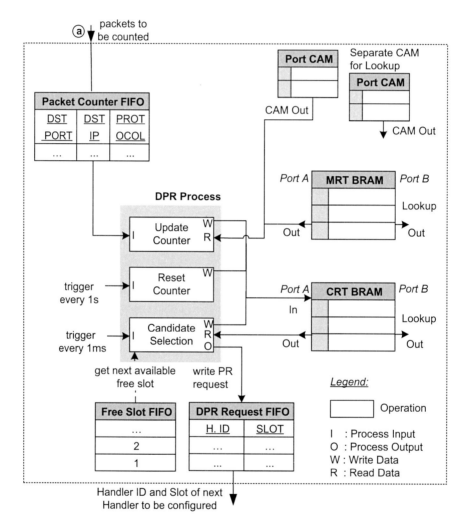

Figure 5.9: Implementation architecture of the adaption engine
©2011 Springer Berlin Heidelberg. Reprinted (modified) from [Müh11a].

cannot handle the request, there are other possible schemes that could be implemented to support application demands. Two of them are discussed in the following text. The first option is to implement an additional queue that holds packets for which a Handler exists in the matching table, but which is not yet online. These packets would then be fed again into the architecture after a set time interval (e.g., 5 ms), and if a Handler is then online, the packet could be delivered. This behavior imitates TCP retransmission internally, but with a shorter frequency. The drawback for this approach is that the queue can fill up quickly, if a lot of packets for different Handlers arrive at the system simultaneously. Furthermore, the internal retransmission of packets increases the load on the internal data path, possibly affecting the overall throughput. As an extension to this approach, a more refined scheduling algorithms could be implemented that would rearrange the queue. E.g., Dynacore [Alb06] uses this approach to direct packets with no suitable hardware accelerator to an intermediate queue for holding.

A second option could utilize the Slot buffers to temporarily store each arriving packet that does not have an on-line Handler. This could be implemented by returning as part of the Slot lookup the value of the Slot into which a currently off-line Handler will be deployed. The packets could then be routed directly to the Slot buffer and wait there until the Handler becomes available. However, this requires a reconfiguration candidate selection run for each arriving packet that has no active Handler, which will affect the overall performance if very heterogeneous traffic arrives. To ensure that only the correct Handler processes the message, a new ICH field should be used to identify the target Handler ID. The drawback of this scheme is, that in the worst case the packet throughput could be reduced to only a few 1000 packets / second, depending on the speed of the reconfiguration process (see Section 7.2.4).

For the honeypot use-cases, the loss of a limited number of packets was deemed more acceptable than the drawbacks of the alternative methods presented here.

5.3.6 Transferring Bitstream Data

The bitstream data for each Handler-Slot combination is transferred to the external memory using the management interface. NetStage trusts the management network and does not perform security checks on the incoming bitstreams. If necessary, encryption solutions using, e.g., AES [Zei05, Hor08], could easily be integrated in the flow.

The bitstream data is divided into multiple single packets using custom scripts on the administration PC and transferred to the NetStage system including a header that contains the destination memory address. The NetStage management module then simply writes the received data to the memory.

5.4 Chapter Summary

This chapter presented details on the application layer of the NetStage platform. Applications are built using so-called Handlers. A single application can consist of one or multiple Handlers, that are all connected to the communication core. For that purpose, a unified interface has been defined that describes the message-based data exchange between Handlers and the core.

To simplify application development, a real-time simulation environment is provided that offers a virtual network interface. This virtual interface connects the HDL simulation with live network traffic and allows to easily verify new Handler developments.

For scaling the platform, NetStage supports dynamic partial reconfiguration of Handlers. In this manner, many Handlers can reside in an offline storage and will be swapped-in on demand (e.g., to adapt to the nature of the network traffic). This is especially useful for the honeypot application, as more vulnerability emulation Handlers than would fit on a single FPGA can be held available for long-term experiments without manual intervention.

6 Malacoda: Compiling Honeypot Applications on NetStage

As described in the introduction, the honeypot scenario has been chosen as a use-case to demonstrate the NetStage platform capabilities. Honeypot traffic consists of multiple requests from different clients accessing multiple services. Furthermore, the security provided by the dedicated hardware is of high relevance here. The resulting hardware-based honeypot has been called "MalCoBox".

While the direct implementation of Handlers for the MalCoBox in hardware programming languages, such as VHDL or Verilog, is possible by experienced developers, it is generally too complex and time-consuming for frequent Handler updates. Furthermore, platform access is limited to developers that are familiar with hardware design. This is doubly detrimental for the MalCoBox use-case: In general, network security researchers will not have the required hardware design experience. However, with the quickly changing attack landscape on the Internet, new vulnerability emulations must be created quickly to keep up.

Because the difficulty of FPGA programming is not only a problem for the MalCoBox, but is instead a general issue for the general usage of FPGAs, research on alternative ways to simplify FPGA programming is a major focus in the research community (see also Section 3.4). This ranges from generic solutions aiming at compiling, e.g., C code to FPGA hardware modules [Gad07, Xil12d] up to specific solutions for, e.g., packet processing [Att11, Bre09] or router development [Rub10]. Many of these solutions are achieving good results when focusing on particular problems (by introducing a Domain Specific Language, DSL).

Such a DSL has advantage for both, the programmer and the compiler. A DSL allows the programmer to describe a specific problem in his domain (results from Brebner show, that a domain-specific language achieves significantly higher efficiency than a general-purpose language [Bre09]), while the compiler on the other side can create extremely efficient hardware circuits due to the fact that the semantics of target applications is known in greater detail. Furthermore, by using a DSL, the intricacies of a hardware platform can be hidden perfectly from the programmer (e.g., by considering inevitable hardware restrictions implicitly in the language definition).

To open the MalCoBox to a broad range of users, a new DSL called Malacoda has been developed, that allows network engineers to describe vulnerability emulations using programming constructs from their traditional domain, while the compiler can generate high-performance Vulnerability Emulation Handlers (VEH) according to the micro-architectural model required by NetStage. As VEHs have a regular structure (see Section 5.1), a suitable compiler could be realized efficiently. The initial design of the language [Müh11b] has been discussed with subject experts, resulting in the description of Malacoda that is presented here [Müh12b]. Details of the language and the compiler are presented in the following sections, partly repeating published material from [Müh12b].

The language and the compiler presented here should be seen as an initial proof of concept to demonstrate the seamless integration of compiled service modules with the platform. Even though its syntax looks familiar (Perl-like), Malacoda has not been designed as a new general-purpose programming language, but instead as a high-level approach focused especially on the honeypot domain. Before starting with the details of Malacoda, the following Section gives a short introduction into DSL development in general.

6.1 Domain Specific Languages

A short and concise description of a DSL is given by Mernik et al. [Mer05]: 'domain specific languages trade generality for expressiveness in a limited domain'. Such languages provide notations and constructs targeting a specific application domain, which reduces the programming expertise needed to describe specific solutions compared to a General Purpose Language (GPL). This leads to increased productivity and opens up the application domain to a larger group of developers. Today, many DSLs are widely used in the computer science domain, such as SQL (Databases), MATLAB (numerical calculations) or HTML (Web pages).

[Mer05] and [Gho11] list the following advantages of developing a DSL:

- Appropriate domain-specific notations are usually beyond the limited user-definable operator notation offered by GPLs. This supports concise and expressive descriptions close to the problem domain, resulting in an overall gain of productivity.

- Analysis, verification, optimization, parallelization, and transformation of DSL constructs are much easier than relying on GPL source code patterns (which are often too complex or not well defined). Furthermore, domain-knowledge can be validated and reused more easily.

- Unlike GPLs, DSLs can be non-executable. DSLs can therefore focus on the description and do not need to deal with execution. Accordingly, DSL 'programs' are often properly called specifications, definitions or descriptions.

However, there are also some disadvantages that come with DSLs [Gho11]:

- Language design is a complex task that requires significant resources. Developing a new DSL is not trivial.

- DSLs can lack tool support, e.g., for debugging and profiling. Solutions for these tasks might need to be developed in addition to the DSL core compiler.

Mernik et al. [Mer05] provide a guideline for four different phases that are usually required when developing a new DSL, which can be summarized as:

Decision: As developing a new DSL may require significant effort, the decision needs to be carefully made. However, the potential ROI is often difficult to predict. E.g., a very good reason for a DSL is an already existing domain-specific notation as an enabler for users with less programming expertise [Sut04]. Furthermore, better support for domain-specific analysis or optimization can be used to make a case for a new DSL.

Analysis: In-depth domain knowledge for the development of a new DSL can be gathered by, e.g, reading technical documents, interviewing domain experts, or inspecting existing GPL source code for that domain. In practice, this process will often be performed informally, e.g., by collecting information until a good understanding has been reached. Formal processes rarely used in this context.

Design: Based on the domain knowledge, the DSL can be designed. There are two fundamental options: inventing a new language or exploiting an existing GPL or DSL. Basing a DSL on an existing language is especially advantageous if the users are already familiar with this language. When designing a DSL from scratch, well known GPL design criteria should be followed (readability, simplicity, etc.). However, established notations of the domain, which should be preserved, may interfere with these principles [Wil04]. This could lead to domain-specific languages that are viewed as less efficiently readable from a programmer's perspective.

Implementation: Multiple options are available for actual implementation of the DSL: Interpreter, Compiler, Preprocessor, Embedding, Extensible compiler, etc. An interpreter takes DSL constructs and interprets them using a standard fetch-decode-execute cycle. The code is thus translated on-the-fly. A compiler translates the DSL description to a base language that can be efficiently executed on the target system (e.g., assembler code for a processor). Implementing an interpreter is easier than building a compiler, but at the cost of a slower execution performance. A preprocessor translates DSL constructs into an existing language, where existing tools (e.g., a compiler) can be used to further process the program. Such an approach is useful, if the target system is very complex to program and writing a new compiler is extremely difficult. The embedding approach embeds DSL constructs in an existing GPL by defining new abstract data types and operators, while extending an existing GPL compiler/interpreter means that it understands DSL constructs directly.

6.2 DSL Decision and Domain Analysis

The two major goals that should be achieved with a high-level compiler on the MalCoBox are:

- Open the MalCoBox for security and network engineers without hardware design expertise.

- Provide an efficient way for hardware-experienced developers to quickly generate service module templates for the NetStage platform, that can be easily modified if the DSL capabilities do not suffice for specialized applications.

These goals can be well supported by a DSL. By using domain-specific notations, the basic activity of a VEH can be represented using a compact representation. Fundamentally, VEHs always perform the same execution cycle: read a packet, process it, and generate a response. There is no benefit from using a GPL here, as the Handler hardware does not support arbitrary functionality at the highest processing speeds.

When searching for comparable DSLs, the first place to look is the domain of software honeypots. However, service emulations for current software honeypots are written in the language of the core honeypot system (e.g., C++ for Nepenthes [Bae06]) or can be included using other languages (e.g., Perl for HoneyD [Pro04b] or Python for Dionaea [Dio11]). No single language is in widespread use in the honeypot domain. On the other hand, GPLs are not well suited to be compiled to hardware for this purpose (see also Section 3.4.1). A GPL would need to be highly restricted with regard to language features and coding style to allow efficient hardware generation. Taking one of these GPLs and using it as the base language for the MalCoBox compiler, is thus not a preferred solution.

When widening the search from GPLs to DSLs for general network processing on FPGAs, G [Bre09] might be a potential candidate (see also Section 3.4.2). G is a general-purpose language, but specialized for packet-header processing. It allows users to flexibly specify the packet format (fields and positions) and conditional rules for modifying these fields depending on packet contents. The programs are then compiled into hardware units, which can be used on an FPGA. However, G lacks facilities such as regular expression handling and extended support for protocols above the level of processing individual incoming packets. Furthermore, the syntax of G is very close to packet header processing, which does not perfectly fit the domain of the honeypot. This also holds true for PPL [Mye06] or PacketC [Dun09], other languages targeted for packet processing. Such generic solutions are too far removed from the honeypot use-case to allow a concise description of the honeypot activity.

As none of the existing solutions fit or could be easily extended to fit, the decision was made to develop a DSL for the MalCoBox from scratch.

6.2.1 Domain Analysis

A good source for collecting the domain knowledge required to design a proper DSL are the available software honeypots and their emulation scripts. Such honeypot service emulations mostly operate on a per-packet basis. Incoming requests for services are evaluated and matched against a set of rules that decide which response template is used to generate a response packet. These response templates contain packet data that would have been sent if the emulation would have been a real service. Sometimes it is useful to add connection-specific data to the response, such as a dynamic session ID. Often such dynamic data can be extracted from the request and directly inserted into the response. Extensive computations are very rare for low- to medium-interaction honeypots that are targeted here, except for cryptographic schemes as part of an emulated protocol. Listing 6.1 shows an example from Nepenthes [Bae06] that implements an SMB interaction [Mic09].

This is a good example showing the dialog-based approach that is followed by most of the service emulations. In that case, the packet is inspected for a valid SMB session initialization header and if this has been detected in a packet, the next state is reached (which will be evaluated when the next packet arrives). If the next packet is a SMB negotiate packet, then the SMB initialization sequence has been performed successfully. In that case, a corresponding log message is sent to the administration station.

For this work, domain analysis has been performed informally by going through a wide range of implementations of software honeypot services. Based on this analysis, the following key operations have been extracted that are commonly used by the majority of implementations. These operations are [Müh12b]:

- Describe a sequence of steps (states) that reflect the communication session.

- Evaluate the incoming request packet and craft a proper response packet by using static template data and parts from the request packet based on certain rules.

- Notify an administration station about certain steps that have been reached.

Malacoda will therefore focus on providing these operations, allowing the implementation of a wide range of current service emulations. All operations need to be designed to act on network packets, including sub functions for accessing single bytes or bits of the packet. Furthermore, the language should be open for optional future extensions, such as more complex computations, which are not in the focus for the current version.

Listing 6.1: Sample Nepenthes Script [Bae06]

```
ConsumeLevel SMBNameDialogue::incomingData(Message *msg)
{
  ...
  switch (m_State)
  {
  case SMBName_NULL:
    {
      char *buffer = (char *)m_Buffer->getData();
      smb_header *sh = (smb_header *)buffer;
      ...
      if (sh->m_RequestType == 0x81)
      {
        m_State = SMBName_NEGOTIATE;
        ...
      }
    }
    break;

  case SMBName_NEGOTIATE:
    if ( m_Buffer->getSize()== sizeof(smb_negotiate_req0) &&
       memcmp(m_Buffer->getData(),smb_negotiate_req0,sizeof(
          smb_negotiate_req0)) == 0 )
    {
      logSpam("SMB Negotiate request %i\n", m_Buffer->
          getSize());
      msg->getResponder()->doRespond(smb_negotiate_reply0,
          sizeof(smb_negotiate_reply0));
      m_State = SMBName_DONE;
      ...
    }
    break;

  case SMBName_DONE:
    break;
  }

  return CL_ASSIGN;
}
```

6.3 Language Design

As scripting languages such as Perl are often used in the network security domain, the syntax of Malacoda is inspired by Perl [Per12]. Malacoda uses basic constructs from Perl and adds domain-specific constructs to allow for easy description of the operations summarized above.

The syntax has been defined using an ANTLR grammatical representation. This formal approach has the advantage, that a tool [Par07] exists which automatically generates a parser framework (including syntax checks etc.). However, please note that the Malacoda language currently allows to describe a wider range of expressions (especially in terms of regular expression matching and variable handling) than the current prototype compiler is able to generate code for. The functionality actually implemented is explicitly described in the following sections.

Table 6.1: Malacoda Commands

addresponse(*SOURCE*)	Append a given byte sequence to the response packet buffer.
addresponse(**file:***STRING*)	Send a given byte sequence defined at compile-time in an external file (useful for larger responses that would make the Malacoda program hard to read if embedded into the Malacoda source)
log (*SOURCE*)	Send log packet with the given byte sequence to management interface.
replace(*s*, *SOURCE*)	Replace a single byte or a byte sequence of the response packet with the value given by *SOURCE* starting at index *s*.
close	Send a close connection notification with this response packet to the client (only available for TCP connections).

6.3.1 Syntax

Listing 6.2 shows a sample Malacoda description emulating a simple Telnet login into a shell, accepting any user / password combination. A Malacoda description [Müh12b] starts with a name, followed by an optional section to define state variables and configuration settings. The keyword **dialog** begins the activity description. A dialog contains a sequence of interactions with the client, including, e.g., the response packets sent back to the client based on the incoming request packets. In that fashion, a dialog represents the communication pattern with the client.

Each stage of the dialog is identified by a state name that is tracked automatically on a per-connection basis using the global application state service of the NetStage core (see Section 4.2.6) and set using the reserved variable $state. The default state DEFAULT indicates the initial state, that is always accessed when a new connection is established or no state information is available (e.g., for UDP connections).

For each dialog step, there is a block of multiple commands that are executed when a new packet has arrived. A new state name identifies the start of a new block and implicitly ends the previous one. All commands in between belong to the previous state. If there is no further state name identifier, the state block is implicitly ended by a closing curly bracket that closes the entire dialog.

Table 6.1 lists the commands of Malacoda that are currently allowed inside state action blocks. *SOURCE* can be either a string (of ASCII characters or byte values, expressed by prefixing two hex digits with \\), or a variable name (reserved or user-defined). *SOURCE* can be further narrowed down by selecting particular bytes with $[s, n]$, where s is the start index and n the length. For particular use cases (e.g., setting flags for custom protocols), single bits can be accessed using the syntax $[s][b]$, where s is the index of the byte and b the index of the bit. Furthermore, the command chomp can be used to remove line breaks from a variable assignment.

Note that a response packet may be incrementally constructed with multiple commands. It will only be sent once all of a state's actions have been processed. The content of the incoming request packet can be accessed using the reserved variable **$INPKG**.

State actions can be executed conditionally using if/elsif/else constructs. The language supports arithmetic expressions and regular expression matching for conditions.

6.3.2 User-Defined Variables

Malacoda allows the explicit storage of state in user-defined variables. These are held in the global application state memory (see Section 4.2.6) for the duration of the entire session. Variables store unstructured byte sequences that are interpreted in context of their current operator. However, they can be declared differently

Listing 6.2: Sample Malacoda description (Telnet emulation)

```
// Emulate login to a root shell
TELNET_VEH {

  //define variables
  dynamic $loginname[14];

  // stateful dialog description
  dialog {

    //initial state
    DEFAULT:
      addresponse("Connected to localhost.localdomain\n");
      addresponse("login:");
      $state = LOGINWAIT;

    LOGINWAIT:
      // store login name to variable
      $loginname = chomp($INPKG);
      addresponse("password:");
      $state = PASSWORDWAIT;

    PASSWORDWAIT:
      addresponse("[localhost]#");
      $state = SHELL;
      log("TELNET: Login");

    SHELL:
      if($INPKG  =~ /^ls/) {
        addresponse("web-password.txt\n");
        addresponse("[localhost]#");

      } elsif($INPKG  =~ /^whoami/) {
        // send back stored login name
        addresponse("$loginname");
        addresponse("\n");
        addresponse("[localhost]#");
      }

      // more commands
      ...
  }
}
```

depending on whether they have a variable length (up to a static upper limit of the maximum GASM cluster size, see Section 4.2.6) or a fixed length (Listing 6.3).

Listing 6.3: User-defined Variables

```
// variable with dynamic length (maximum length in bytes)
dynamic variable1[8];
// variable with fixed length (in bytes)
fixed variable2[4];
```

In the first case, an additional byte of storage is used to track the length of the variable. Longer values assigned during runtime will simply be truncated to the maximum variable length.

Variables can be used to track custom state, user names etc. Note that variables are global for the entire dialog. Furthermore, local modifications of variables inside a state action block will not be visible in the current iteration of the execution loop. The updated variable content is only available with the next incoming packet.

6.3.3 Expressions

Malacoda supports arithmetic, regular expression, and comparison operators in expressions. The current version of the language uses unsigned byte sequences as the fundamental data type. Data is interpreted in context of the current operator, e.g., incrementing a value and then interpreting it as a string is a valid operation.

The syntax of regular expressions is based on the Perl compatible regular expressions (PCRE), which is the de-facto standard for network applications. However, please note that the current compiler prototype does not support any arithmetic expressions.

6.3.4 Log Messages

Log messages can be assembled using the given content *SOURCE*. However, only fixed byte sequences are supported by the current prototype compiler. Reference information (e.g., IP addresses) is automatically added to each log message and does not need to be specified here.

6.3.5 Malacoda Examples

Listing 6.4 shows a Malacoda code fragment for emulating a DNS server that responds to queries for a particular domain with a predefined response (read from a file at compile time) and logs a message to the management station at run-time. As DNS is stateless, the dialog consists only of the **DEFAULT** state.

Listing 6.4: Excerpt from DNS server emulation

```
DNS_VEH {
   dialog {
     DEFAULT:

        // check if it is a query
        if ($INPKG[3][7] = 0) {

          // check for DNS request name (skip 12 header bytes)
          if ($INPKG =~ /^.{12}\\03www\\08malcobox\\02de/) {

            // copy the entire request packet to the response
            addresponse($INPKG);

            // keep the ID but replace header
            replace(3,"\\81\\80");
            replace(7,"\\00\\01\\00\\02\\00\\02");

            // add the IP address response
            addresponse(file:"dns_response.txt");

            // notify management station
            log("DNS Packet with Request for www.malcobox.de")
               ;

          } else {

            ...

          }
        }
      }
   }
}
```

Another Malacoda example is shown in Listing 6.5. This Handler emulates a web server by sending an HTML page (e.g., a login page). The response data is defined in two external files (one for the HTTP header and one for the page content).

Listing 6.5: Excerpt from web server emulation

```
HTTP_VEH {
  dialog {

    // stateless VEH, only initial state
    DEFAULT:

      // send HTML response if request for particular URL
      if ($INPKG =~ /GET \\2elogin.html/) {
        addresponse(file:"login-http-header.txt");
        addresponse(file:"login-html-content.txt");
      }

      // send other responses (e.g., redirects)
      ...

  }
}
```

6.4 Compiler Implementation

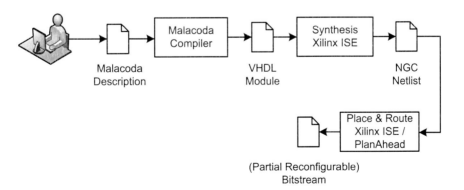

Figure 6.1: Tool flow for generating hardware from Malacoda programs
©2012 ACM. Reprinted (modified) from [Müh12b].

Applications developed in a DSL can be compiled to hardware in multiple ways. The most common way is to translate DSL descriptions into a hardware description language representation, e.g. VHDL or Verilog, which can then be synthesized by the available standard FPGA vendor tools into dedicated hardware (see Figure 6.1). According to the definition of Mernik et al. [Mer05], this approach corresponds to the preprocessor pattern. The Malacoda compiler also follows this way [Müh12b].

Alternatively, there are tools that allow to implement a custom compile flow, such as HMFlow proposed in [Lav11] for rapid prototyping using hard macros. However, such a solution requires proper macro definition to reach high speeds, which in turn requires specific knowledge of the FPGA basics and is much more complex and less portable to different FPGA families.

As an alternative to the generation of dedicated hardware, micro-programmable systems have been proposed that consist of a fixed hardware control module, whose operation can be controlled using a dynamic configuration. However, the flexibility of this approach is bounded by the implementation of the control module. Such a solution has been, e.g., proposed by Kulkarni and Brebner by using a micro-coded data path engine with the example of NAT translation [Kul06].

Other work considers the implementation of a controller engine running descriptions for HFT applications [Lit11] directly without hardware compilation. But the flexibility of such microcode engines is often too limited (e.g., in terms of the required flexible regular expression support) for the honeypot application to achieve the 10 Gbit/s processing throughput required here.

Finally, the compilation from Malacoda into VHDL has several advantages here:

- Complex hardware compilation of the VHDL modules can be performed using the well-established existing toolchains provided by the FPGA vendors.

- Experienced developers can easily add custom functionality to the well-structured generated VHDL modules.

- Simulation of service emulations can be performed using standard VHDL simulation tools.

The generated VHDL modules can either be used to generate a partial reconfigurable bitstream or merged with the NetStage core into a monolithic design. All these steps can be automated, so that an integrated tool can perform all these steps without user interaction.

The VHDL code that is generated by the Malacoda compiler follows the basic Handler module described in Section 5.1. Packets are read from a buffer, processed and responses are written back to a buffer. A central state machine controls the entire operation. The micro-architecture of the compiler (see next section) focuses on generating high-speed implementations of the Handlers that can process data at 10 Gbit/s. The compiler should be able to generate Handler modules that achieve native (or near-native) performance compared to Handlers built manually.

To achieve this goal, the internal architecture of a compiled Handler (see Section 6.4.1) aims at assembling one 16B output word in one cycle wherever possible. In that fashion, it utilizes the entire bandwidth of the data path. If data for an output word needs to be fetched from multiple data sources, dedicated logic for assembling that particular output word is created from each Malacoda statement. Such a wordwise, dataflow-oriented processing is common for compiling high-speed packet processing applications [Sov09, Att11], as it follows the efficient streaming approach supported by FPGAs [Att06, Neu08].

6.4.1 Handler Microarchitecture

Figure 6.2 shows the run-time execution cycle of a compiled Handler. This is the basic template that is used for all Handlers generated by the Malacoda compiler. Initially, the created hardware evaluates all conditions in the Handler in parallel. This is done on a per-word basis while the packet data is arriving from the input buffer. Each condition has a corresponding flag register that is set based on the matching of the condition. Furthermore, the variable values are extracted from the global application state data field of the ICH (if state data is used for this Handler).

Next, the condition match registers are evaluated in program order. To this end, the conditional control structures of each state block are mapped to a single FSM state that is executed based on the current application session state value. The separation into multiple FSM states instead of evaluating the condition in a large conditional block is done to reduce the complexity of a single FSM state, it does not add additional execution cycles. The conditional evaluation predicates the execution of the state's actions, which are then executed sequentially.

The execution inside each code block occurs in the following order: First, the variables are assigned, then a response packet is generated, and finally a log message is built. Note that there can be multiple variable assignments, but currently only one response packet generation and one log message generation block inside a state execution cycle are supported.

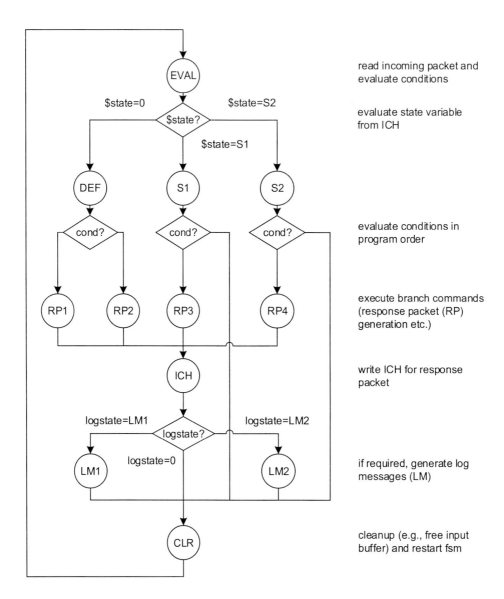

read incoming packet and
evaluate conditions

evaluate state variable
from ICH

evaluate conditions in
program order

execute branch commands
(response packet (RP)
generation etc.)

write ICH for response
packet

if required, generate log
messages (LM)

cleanup (e.g., free input
buffer) and restart fsm

Figure 6.2: Execution cycle of a compiled Handler
©2014 IEEE. Reprinted (modified) from [Müh14].

The specific states for generating a response packet only *create* the actual payload of the NetStage message. The ICH is *written* in a shared state. The ICH of the request message is copied and control information (e.g., payload size, close connection) is used to update the values of particular fields. After the ICH has been written, the next NetStage stage is notified of the completion of the packet, which can then be prepared for transmission. As one of the last steps, an optional log packet could be generated. Finally, a cleanup state assures that the packet is cleared from the input buffer. For each new arriving packet, the Handler-internal FSM process starts from the beginning.

6.4.2 Compile Flow

Due to the highly specialized nature of Malacoda, much of the complexity of conventional high-level language compilers [Koc10] can be avoided. The compile flow itself is shown in Figure 6.3 [Müh12b]. The basic compiler template is using Java code automatically generated by the ANTLR v3 compiler-construction tool [Par07]. It not only generated the lexer and parser from a high-level description, but also the Abstract Syntax Tree (AST) representation.

During the semantic analysis pass, the AST generated by the parser is initially traversed to build a symbol table of states and variables. Furthermore, conditions (including regular expressions), response packets, log packets, and variable assignments are collected and stored in central tables. This is done to later aggregate command groups (see below). Additionally, the total number of static bytes used by response packets is stored in a global variable for later use by the code optimization logic.

Also, the analysis run determines the basic blocks and their control predicates, storing this data by annotating the AST. This results in a compact representation of the original syntax tree. The annotated tree structure is shown in Figure 6.4.

The top VEH node has the individual state nodes as children. Each state node corresponds to a state block of the VEH dialog description. The state nodes can contain either if/elsif/else constructs or command group nodes as children. If/elsif nodes contain a reference to their related condition and can have itself command group nodes or another if/elsif/else block as children.

A *command group* consists of all single occurrences of a particular command within a block. E.g., it will be often the case that there are multiple single **addresponse** commands to increase the readability of the VEH description. As there will be only a single response packet per execution cycle, all single commands that belong together will be aggregated.

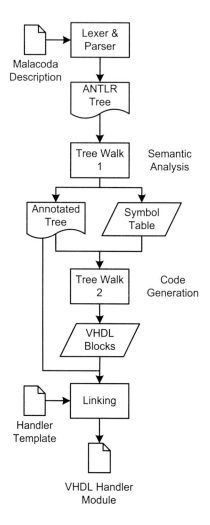

Figure 6.3: Malacoda compiler organization
©2012 ACM. Reprinted from [Müh12b].

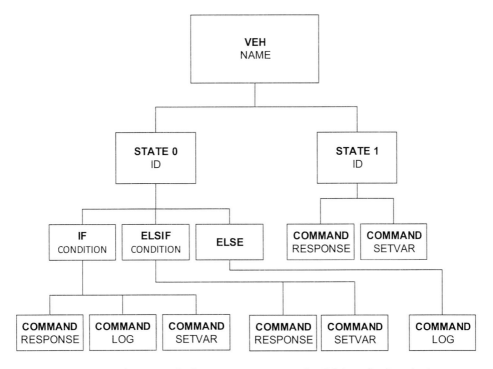

Figure 6.4: Annotated abstract syntax tree for Malacoda descriptions

The information of the annotated AST is exploited in the code generation pass, which expands a pre-defined Handler template in VHDL by replacing placeholders with the actual signal declarations and output assignments. The template already contains the buffered interface to the platform's network core and a skeleton FSM for receiving and sending messages, which is then extended with the Handler-specific processing. Figure 6.5 sketches the different parts of this template.

Code generation proceeds as follows: First, the static contents of response and log packets are inserted into the template. Then, all individual conditions are translated into their underlying hardware (e.g., regular expression matchers, simple comparators), computing boolean signals. A condition evaluation block combines these predicates into the more complex expressions that actually control the execution of the basic blocks within a state (also considering intra-state control flow, if any). The contents of the basic blocks are assembled from code block templates for the different possible statements. These include packet

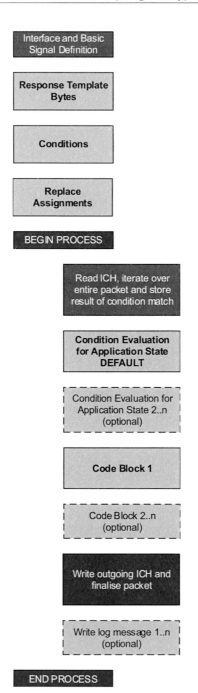

Figure 6.5: Handler code template
©2014 IEEE. Reprinted (modified) from [Müh14].

generation (response and log) and variable assignments. The statements to perform replace operations are handled separately to increase the flexibility. The following paragraphs describe in more detail the content of the different blocks.

6.4.3 Conditions

The single conditions that have been collected in the condition table during semantic analysis are separately translated into VHDL conditions. Basic conditions that match on a fixed byte sequence at a given location inside the request packet can be mapped directly to VHDL code. Currently, the compiler supports basic byte comparisons on equality, whereas the data to compare can be at arbitrary positions in the packet.

As NetStage messages always start with the data payload at a fixed location, the offset for the match operations can be easily calculated. As the packet is read word by word, word-wrapping conditions are generated using intermediate registers to store if parts of the condition already matched a previous word. Comparison and logical operators are directly converted to VHDL equivalents.

E.g., the Malacoda condition

```
if ($INPKG[0,1] = "\\81")
```

will be compiled directly into the following VHDL code:

```
condition_1 <= '1' when ((packet_data_in(7 downto 0) = x"81"
    ) AND (numwords = 1 + data_offset)) else '0';
```

The statement matches the first byte of the input packet. **numwords** is the counter for the current word of the input packet, which is stored in **packet_data_in**. **data_offset** is used to handle variable application state data regions lengths in the ICH.

If, e.g., a dialog has now been defined as:

```
dialog {

  DEFAULT:
    if ($INPKG =~ /^ls/) {
      addresponse("passwd.txt\n");
```

```
      addresponse ("[localhost]#");
      $state = LSSEND;
   } else {
      addresponse (file:"commandnotfound.txt");
   }

   LSSEND:
      if ($INPKG =~ /^cat passwd.txt/) {
         addresponse (file:"passwd.txt");
      } else {
         addresponse (file:"commandnotfound.txt");
      }
}
```

the corresponding VHDL code for the FSM state blocks will be:

```
when state_0 =>
   if (condition_1_matched = '1') then
      mainstate <= state_1;
   else
      mainstate <= state_2;
   end if;

when state_3 =>
   if (condition_2_matched = '1') then
      mainstate <= state_4;
   else
      mainstate <= state_5;
   end if;

when state_1 =>
   romindex <= returnpacket_1_start;
   mainstate <= return_packet_1;

when state_2 =>
   romindex <= returnpacket_2_start;
   mainstate <= return_packet_2;

   ...
```

The states (**state_0** etc.) indicate main FSM states for the VHDL process that manages the handler execution cycle (see above). **state_0** and **state_3** are states that depend on the setting of the state variable **$state**. After reading the input packet for evaluating the conditions, the main FSM continues with either of them. Inside these state blocks, conditional execution will be handled directly following the Malacoda definition. If, e.g., condition 1 has matched, then the FSM will continue with **state_1**, which contains the information to continue with sending response packet 1. The reason for separating this into multiple states is to keep the conditional evaluation tree small (to reduce the signal delay), if there are multiple nested conditions in a Malacoda description. The numbers referring to conditions, states, response packets etc. are automatically assigned by the compiler during the semantic analysis pass.

6.4.4 Regular Expression Matching

Conditions that go beyond the basic matching described above can be implemented using regular expressions. Regular expressions are extremely relevant in most networking applications. For example, they are excessively used for network intrusion detection systems such as Snort. But regular expressions are extremely costly when run on GPPs, especially if multiple regular expressions are matched against the same input. Here, the massive parallelism available on FPGAs allows major performance improvements especially for large scale setups with thousands of rules that can be matched.

Efficient implementations are a particular focus of the research community (see Section 3.1.2). However, while these solutions provide excellent results for large-scale setups, they are a bit oversized for the Malacoda compiler, as each MalCoBox Handler contains only a small number of regular expressions (ca. 5-10). Furthermore, Handlers should be kept simple, while the deep pipelining and other techniques employed by the large scale regular expression compilers should be avoided here to achieve a tiny footprint.

Therefore, the regular expression compiler for the MalCoBox uses simple FSMs with maximally parallel comparators (16 input characters) to ensure that a result is available immediately after a 16B input word has been read from the buffer (adhering to the general micro-architecture). The compiler generates such a parallel shift-and-compare [Bak06] matching engine for each regular expression in a Malacoda program, where no absolute position information is given. This results in a high overall regular expression throughput and supports the effort to achieve at least 10 Gbit/s throughput in the compiled Handler.

However, this is a resource intensive approach that only works well if there are no more than a couple of regular expressions per Handler (which is in fact true for the honeypot). It could be worthwhile to integrate more refined matching architectures, that also support multiple input characters, e.g. [Yam08], for other applications if they expect a higher number of regular expressions.

In the current implementation, the regular expression support is limited to byte (e.g., characters or strings) comparisons that can be further constrained by giving an exact position in the packet (either at the beginning, somewhere in the middle or at the end). They are preferably used to compare strings that could not be that well represented using basic conditions.

E.g., the Malacoda condition described using the regular expression

```
if($INPKG =~ /^GET \\2fwebmail.php\\3finclude\\3dlogin/)
```

will be compiled into the following VHDL code:

```
condition_1 <= '1' when ((packet_data_in(127 downto 0) = x"
    7068702e6c69616d6265772f20544547") AND (numwords = 1 +
    data_offset)) else '0';
condition_2 <= '1' when ((packet_data_in(111 downto 0) = x"6
    e69676f6c3d6564756c636e693f") AND (numwords = 2 +
    data_offset)) else '0';
condition_3 <= '1' when ((condition_1_matched = '1') AND (
    condition_2_matched = '1')) else '0';
```

In this case, the condition will be be assembled using three separate statements. Statement 1 matches the first 16 Bytes of the match string, while statement 2 matches the second part. Statement 3 combines the two statements (**condition_1_matched** is the corresponding register for the **condition_1** match signal) and holds the final result. It is actually the one that will be evaluated when executing the FSM state block that implements the conditions matching in program order for that particular state.

6.4.5 Variables

As already described, variables are defined to be global for a single connection and exist during the entire dialog spanning multiple packets. Variables are implemented inside the compiled Handler by using pairs of two registers for each variable. One holds the original value of the variable retrieved from GASM and transported in the ICH, and the other register holds the new value to be written back to the GASM. Initially, both registers are set to the incoming value of the

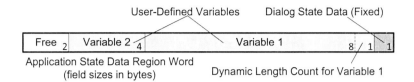

Figure 6.6: Variable allocation in Global Application State Memory

ICH. If a variable remains unchanged, its previous value is written back to the ICH.

For the implementation it has been defined that read accesses to a variable are always served by the register holding the original value, even as the other register already contains a newer value written during the current execution cycle. This is fine for the honeypot application and allows to easily schedule variable handling at the beginning of a state execution cycle (see Section 6.4.1). Note that this implies that incremental changes of variables are not possible inside an execution cycle, but of course during an entire dialog. If necessary, more expressive schemes such as Ephemeral History Registers (EHR) [Ros04] could be implemented later.

As the GASM provides only a region of raw data words, the compiler has to define the actual allocation of variables inside the application state data region of the ICH. The compiler allocates space for the reserved internal (currently only **$state**), as well as for the user-defined variables (Figure 6.6) [Müh12b]. Dynamic variables occupy their maximum length plus a byte tracking their current length, static variables always have a fixed size. The explicit storing of length information for dynamic variables is important so that the wiring networks used in packet construction can determine the position where to append further data to the packet (see next section).

Currently, the compiler supports the use of variables only as part of an **addresponse** command. Assignments to variables can be based on fixed positions inside a packet or by using the entire packet. Optionally, closing newline characters can be removed by using the **chomp** command.

E.g., the Malacoda variable assignment

```
fixed myvariable[4];

dialog {
  ...
  myvariable = $INPKG[0,4];
```

```
    ...
  }
```

will be compiled into the following VHDL code:

```
-- Mapping into Global Application State Memory data region
    of ICH
alias myvariable: std_logic_vector(31 downto 0) is
    state_data(0)(47 downto 16);
alias myvariable_new: std_logic_vector(31 downto 0) is
    state_data_new(0)(47 downto 16);

-- Separate state for variable assignment
when variable_0 =>
  ...
  myvariable_new <= reg_packet_data_in(31 downto 0);
  state_write <= '1';
  ...
```

First, the compiler generates the mapping into the global application state memory data region of the ICH, both for the register holding the original incoming values and for the register holding new values. Then a separate state is generated in the main FSM that manages the variable assignment by copying the declared range of input data to the variable register. As said before, variable assignment is scheduled before response packet generation, so that after variable assignment in this example the FSM continues with the response packet generation state **return_packet_3**.

6.4.6 Response Packets

NetStage Handlers aim for generating response packets at speeds of 10 Gbit/s and more. To achieve this goal without introducing large pipelines and multiple processing engines per Handler, the output packet generation process should generate a valid 16B data word of the packet within each cycle. Note that due to the ring buffer output structure this does not imply that the words need to be processed in the order of their appearance in the output buffer. If other access orderings are more suitable, these can be chosen. If the Handler executes at the native core speed of 156.25 MHz, this results in a maximum theoretical throughput of 20 Gbit/s. However 'dead' cycles (because they are not generating a packet output) that might be required in some Handlers to process variables or log messages lower the overall throughput.

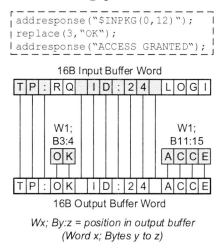

Figure 6.7 code block:
```
addresponse("$INPKG(0,12)");
replace(3,"OK");
addresponse("ACCESS GRANTED");
```

16B Input Buffer Word

16B Output Buffer Word

Wx; By:z = position in output buffer
(Word x; Bytes y to z)

Figure 6.7: Static output word generation
©2012 ACM. Reprinted (modified) from [Müh12b].

As shown in Figure 6.7, response packets are generated by copying predefined content from stored templates, which is then modified using Malacoda commands (e.g., **replace**) at run-time [Müh12b]. The compiler can implement these template ROMs (which are also 128b wide) either as LUTs (for small templates) or BRAMs (for larger ones). Note the key difference to conventional approaches: the operations in the program are *not* mapped to byte-wise copy-and-select steps when reading a template. Instead, they are turned into a dedicated wiring/logic network that modifies all bytes of a 16B template ROM data word in parallel.

While this is easily implemented for static template operations, where the absolute location in the output packet is known, more effort is required for operations that copy variable-length parts of the input packet or dynamic variables to the response packet. In the latter case, the current offset for edits must be tracked during packet construction and the content will need to be inserted at variable locations (depending on prior variable-length data, see Figure 6.8) [Müh12b]. To be able to achieve this within a single clock cycle, a barrel shifter is used to provide the content pre-shifted by 16 different positions, so that the bus holding the appropriately shifted content can be selected. However, this dynamic editing (if required) adds one additional clock cycle for the entire packet generation process due to a required extra register.

For each response packet a dedicated code block is compiled that executes the required operations (see Figure 6.5). Currently, there is no sharing of template fragments. Replace statements are evaluated separately outside of the state machine to increase the flexibility of the code block fragments. They overrule any other assignment so it makes sense to let them manipulate the packet word after all other operations have been performed. Therefore, they are working on the write packet word pipeline register directly.

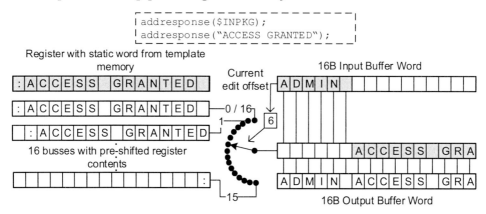

Figure 6.8: Dynamic output word generation
©2012 ACM. Reprinted (modified) from [Müh12b]

6.4.7 Log Packets

Log packets to the management console are generated in a similar fashion, but they will always be extended with the IP source and destination address/port information of the original packet. In practice, log messages mostly serve to identify a packet in an external dump file, with further analysis being performed in the management station.

Note that the target destination address for the management station is fixed during compile time and added to the packet when processed by the management module. The Handler cannot select different targets for log messages. They all go to the same destination machine.

6.4.8 Optimizations

The compiler performs several optimization steps to achieve a better performance in certain situations.

For some protocols (e.g., DNS), the response packet contains much of the data received in the original request packet. The compiler detects this by looking for an `addresponse($INPKG)` command (which copies the input buffer to the output buffer) in the Malacoda program. In that case, dedicated wiring is generated to perform this forwarding in hardware as soon as an input packet arrives. This avoids having to read the entire packet again when building the output packet. Additional static or dynamic data can of course be appended at the end of the copied block. Note that this operation is speculative: If a control condition would actually select a different execution path, the prematurely copied packet is instantly discarded from the output buffer simply by resetting its write pointer [Müh12b].

Typically, LUTs are used to store the packet templates directly. However, for larger packet templates storing them in BRAMs is much more efficient than using LUTs directly. The compiler thus counts the number of template bytes and selects either of the options based on a threshold that has been determinded experimentally (see Section 7.2.2). Note that the BRAM implementation adds an additional cycle to the execution time due to the BRAM data fetch. Furthermore, the current BRAM implementation does only work for static response templates. If a variable is used, the compiler automatically selects LUT as optimization scheme regardless of the template size.

The barrel shifter used for dynamic editing of variable content into packets is a construct that is very costly. It should be used only if absolutely required. The compiler specifically checks for this during analysis and generates the complex hardware only if dynamic commands were used inside the VEH. In their absence, the barrel shifter is not used because the data layout can be statically computed at compile time.

Finally, the handling of state data is only implemented if the Handler actually uses state information. If the Malacoda description contains only the **DEFAULT** state and no other custom variables, the logic to handle state data is omitted to conserve hardware resources.

6.5 Chapter Summary

This chapter presented Malacoda, a domain-specific language for describing honeypot service emulations on a higher level, which can be compiled to high-performance hardware modules. By using a DSL instead of relying on a general purpose programming language, hardware-level restrictions can be embedded in the language definition sparing the user from considering these details when developing new Handlers. Malacoda descriptions allow defining a dialog-based communication session, where incoming packets are answered with properly crafted response packets. These response packets can be generated using packet editing commands such as inserting byte sequences, replacing characters or copying data from the input packet. Conditional execution of commands is supported using regular expressions or basic comparison operators.

The implemented Malacoda compiler generates a VHDL module from a Malacoda description that follows the basic NetStage Handler model as described in Section 5.1. To achieve a high throughput, commands are scheduled for execution immediately when their data is required to generate an output data word. Using this strategy, a full data word will be produced in each clock cycle, thus achieving a high overall performance able to feed the 10G network link of the NetStage base platform. Several optimizations performed by the compiler aim at keeping the overhead of control structures small, depending on the commands actually used in each Malacoda description.

7 Experimental Results

This chapter will focus on the practical results gathered from the NetStage implementation, the Malacoda compiler and the MalCoBox honeypot application. Results presented in this section were obtained on the NetFPGA 10G hardware base. Note that the results presented here use the latest revision of the NetStage platform, as described in this work. Results thus differ from previously published results [Müh10c, Müh11a, Müh12b] on NetStage due to improvements and additions, achieved by continuous development. These refinements have been discussed in the previous Sections.

The results are presented in four categories: Hardware Implementation, Hardware Synthesis, Network Performance and Live Evaluation. Hardware implementation describes the concrete implementation on the test platform that has been used here. Synthesis results focus on FPGA design parameters, such as hardware utilization and clock speed, as well as partial reconfiguration. Performance evaluation shows simulation and practical measurements regarding the system behavior (throughput and latency) under load. Finally, the live evaluation Section gives the results from a long-term run of the hardware honeypot connected to a real high-speed Internet uplink.

7.1 Hardware Implementation

The core platform has been designed to be easily portable between different hardware boards. However, the *practical* implementation on the two available target hardware boards described in Section 2.2 requires board-specific glue logic at the boundaries of the NetStage Core. Before beginning the discussion of the results, the following text covers some of the board specifics. As both the BEE3 and the NetFPGA 10G use Xilinx Virtex 5 series FPGAs, results from both platforms are comparable. The NetFPGA 10G board uses a single TX240 FPGA, while the BEE3 uses a mix of four smaller LX155/SX95 devices.

7.1.1 BEE3 Prototype

The standard NetStage BEE3 FPGA design uses one of the larger LX155T FPGAs and a single 10G network interface for public and management traffic. The separation of network data is achieved by employing multiple different MAC addresses on the single physical link. Later, packet streams are internally split and merged by the interface glue logic.

The physical Ethernet network interface is realized using the Xilinx XAUI [Xil10b] and 10G MAC IP [Xil11b] cores. The external CX4 connectors are driven by RocketI/O transceivers connected to the XAUI block. The 10G MAC provides a Xilinx LocalLink protocol interface that matches the base implementation of the NetStage core, so that it can be attached directly. The DDR2 SDRAM connected to the FPGA is used for DPR and holds Handler bitstreams only (on the BEE3, no external memory is available for core datapath functionality). The required memory interface is implemented using the Xilinx MIG core [Xil10a].

For testing, the network interface is connected to a 10G CX4 port of a dual-XEON Linux server with in total 8 cores and 16 threads (2x XEON E5520, 2.67 GHz, 10GB Ram) running Linux (Fedora Core 14 [Pro10]). This server is used to perform functionality tests, stress-testing and to access the management functionality.

7.1.2 NetFPGA 10G Prototype

The NetFPGA 10G card has been placed in a 1HE Linux server (1x XEON E3-1240, 3,3 GHz, 8GB RAM, OS: Fedora Core 14). The NetStage implementation for NetFPGA 10G uses two of the four available network interfaces, equipped with 10G SFP+ optical transceivers. They are connected to the Linux server, which is also equipped with 10G optical transceivers. In case of the live test with public Internet data, the public interface has been connected to the Internet uplink router, while the administration interface remains connected directly to the server.

In contrast to the BEE3, which provides almost no board-level IP, the NetFPGA 10G platform contains a rich framework of integration modules that provide network connectivity. A physical network interface again consists of the MAC and XAUI combination as on the BEE3, but NetFPGA provides an interface block that encapsulates this as an AXI4 core using the streaming protocol. The AXI4 streaming protocol differs only slightly from LocalLink [Xil11f], so that a wrapper could be developed. The NetStage implementation on the NetFPGA

10G card is based on the NetFPGA Loopback example design [Net11a], which also provides a Microblaze CPU for quick and easy configuration of the SFP+ module parameters.

By default, the NetFPGA provided AXI4 network interface supports different clock speeds for the network interface and the AXI4 streaming port. As this does not match the requirement for a *single* clock throughout the whole main architecture (except for the memory interfaces), the NetFPGA network interface block was modified for implementing NetStage to support a single clock only. This modification resulted in a *major* improvement of the placement and routing results for the design, and were essential for achieving the 156.25 MHz clock speed.

However, with this modification special handling is required for the management interface. As the network clock for each interface is driven by an external clock generator, simply using a single network clock for both interfaces induces the risk for timing issues on the network side. Therefore, the public interface has been selected to be run at the master clock and the management interface is implemented using the original NetFPGA interface block allowing different clock domains. As the management interface contains substantially less logic, it is less susceptible to clock domain-crossing induced timing problems.

The three high-speed external QDR-SSRAMs are used to store global application state data, notification messages and optional bitstreams for DPR (for testing this feature on the NetFPGA board). NetFPGA provides AXI4 cores for the QDR-SRAM memory interface, but as NetStage does not support AXI4 on *internal* interfaces, the memory interface has been implemented using a native interface controller based on a Xilinx application note [Gop10]. In the current setup, the QDR-SRAMs are clocked at 156.56MHz.

7.1.3 Buffer Management

In general, buffer management is an important issue for network systems. The size of the buffers has a significant influence on how a system can deal with varying processing throughputs and local stalls. If the buffers are too small, unnecessary packet loss is induced, while too large buffers can have a negative impact on latency [Get12]. Furthermore, the required high-speed memory is an important cost factor in system design.

7.1.3.1 Buffer Handling

The default NetStage configuration drops input packets when the buffers are full. This scheme is applied for the input pipeline. However, the response path needs a different treatment. E.g., in case of TCP, an incoming packet was already acknowledged when the corresponding outgoing packet is generated (see Section 4.3). If the outgoing packet were dropped because of full buffers, the sender would never send the request again and the connection will get locked in a stall state.

To avoid packet loss inside the architecture, the goal is to guarantee packet transmission on the transmit path. To this end, Handler operation is blocked whenever a send buffer fills up, so that no more packets can be generated, but generated output packets will never get dropped. In turn, if the transmit pipeline fills up, this can lead to packet loss on the receive pipeline, as the Handlers now stop reading from their input buffers, so that they will fill up. However, as long as a packet has not reached a Handler, this packet-loss can be compensated for by external mechanisms (e.g., TCP retransmission).

An additional factor to be aware of is that in the NetStage application domain, the number of bytes that are sent do not necessarily correlate with the number of bytes received (which is different compared to a platform that just monitors network packets flying by). E.g., a Handler emulating a web server can send a 15KB response in reply to a 100B request. If in that case requests would come in at the rate of 10 Gbit/s, a transmit bandwidth of roughly 150 Gbit/s would be required to serve all requests. But as the outgoing bandwidth is also 10 Gbit/s, many requests will be simply dropped even as the input path has enough capacity to receive the packets. Therefore, buffer behavior is also highly dependent on external traffic characteristics, which is a general issue on server systems that will no be discussed here in greater detail.

7.1.3.2 Buffer Sizing

For the store-and-forward approach used in NetStage, an important input parameter for buffer calculation is the expected maximum transfer unit (MTU) of packets that will be stored inside the buffers. This number is dependent on the application scenario. In case of the MalCoBox showcase, possible attack packets arriving from the Internet will have mostly transited wide-area networks. Taking this into account, packet sizes should rarely exceed the Gigabit Ethernet MTU limit of 1500 bytes. Therefore, the current implementation of NetStage assumes an MTU of 1500 bytes and does not handle the larger Jumbo frames that could be used on the 10G network channels.

The NetStage core pipelines are designed for a steady-state throughput of at least 10 Gbit/s for incoming and outgoing packets each. Due to the store-and-forward approach of the Handlers, at least two maximum sized packets need to fit into the buffers (the one that is actually written and the one that is actually read) to assure forwarding of packets without packet loss, which results in a buffer set to at least 3000B. Furthermore, up to 240B should be reserved for the GASM region of the ICH per packet. As the ring-buffer BRAMs have a width of 128b, the next practical value would be 4KB (256 x 128b). However, the Xilinx BRAMs available on the Virtex 5 FPGA require using always four of the basic BRAM blocks to achieve an input width of 128b. This results in a total buffer size of 8KB, which also leaves some headroom for local stalls in the processing pipeline. These could, e.g., happen if large portions of global application state data are requested and can be ameliorated using the larger buffers.

7.2 Hardware Synthesis Results

Synthesis has been performed using either Xilinx ISE (for Handlers and the BEE3), PlanAhead (for DPR) or EDK tools (for NetFPGA 10G) from the Xilinx software suite version 13.3 [Xil11c], depending on the platform requirements. As a common parameter, the entire core design runs at 156.25 MHz (the speed of the network interface) to avoid multiple clock domains for performance reasons.

By configuration, two different design types are supported: static and dynamic. The static version holds all Handler modules together with the NetStage core in a single monolithic design without constraining their placement to particular FPGA partitions. The dynamic version introduces FPGA *partitions* for the Slots, so that Handlers can be replaced using DPR.

The design uses CAMs for multiple purposes (e.g., for the routing lookup). As the previously available Xilinx CAM IP has been replaced with example code starting from ISE Version 13.1 upwards, the design uses the new code from Xilinx [Loc11] for any CAM that is instantiated in the design.

7.2.1 Core Synthesis

Table 7.1 presents synthesis results for the NetFPGA 10G board (Virtex 5 TX240T FPGA) with different configured feature sets. Initially, each design is configured to contain six empty Slots (to measure the core here, not the Handlers). The feature sets are:

a) **w/o SRAM state data and w/o statistics:** This configuration uses internal BRAMs for application state data (128KB) and does not have the statistics module enabled (see Section 5.2.2). This configuration can be executed on the BEE3 too.

b) **with SRAM state and w/o statistics:** This configuration uses the SRAMs on the NetFPGA 10G card to hold GASM data and notification events in external memory, but does not have the statistics module enabled. This configuration can only execute on the NetFPGA 10G card.

c) **with SRAM state data and with statistics:** This configuration uses the SRAMs on the NetFPGA 10G card to hold GASM data and notification events in external memory and has the statistics module enabled. This configuration can only execute on the NetFPGA 10G card.

Table 7.1: Synthesis results for NetStage core components

Module	LUT	Reg. Bits	BRAM	Max. Freq.
(a) w/o SRAM and w/o statistics				
NetStage Core (incl. TCP)	16,296	19,287	107	164 MHz
– Single Slot w/o Content	536	685	4	286 MHz
Management & Routing	3,035	3,613	28	188 MHz
NetStage (6 Slots)	**19,198**	**22,853**	**135**	**164 MHz**
(b) external SRAM enabled and w/o statistics				
NetStage Core (incl. TCP)	17,600	21,539	79	164 MHz
– Single Slot w/o Content	536	685	4	286 MHz
Management & Routing	3,035	3,613	28	188 MHz
NetStage (6 Slots)	**22,500**	**27,563**	**107**	**164 MHz**
(c) external SRAM and statistics enabled				
NetStage Core (incl. TCP)	27,803	27,332	79	164 MHz
– Single Slot w/o Content	938	977	4	286 MHz
Management & Routing	4,128	3,963	28	190 MHz
NetStage (6 Slots)	**33,806**	**33,707**	**107**	**164 MHz**

Without external state data, the NetStage core requires 10% of the LUT resources and 33% of the BRAMs. The high number of BRAMs reflects the several internal buffers that are used to connect the different modules inside the core. In relation to the FPGA capacity, the size of the core infrastructure incl. core, management, slots and routing is comparatively small (12% of LUTs), which leaves sufficient room for Handlers.

Comparing these results with previous results from [Müh11a] and [Müh10c], the increased number of register bits is due to additional registers that have been introduced to assure a stable timing under various conditions on the BEE3 and the NetFPGA hardware.

In configuration b), where external SRAM is used to hold the state information, the overall logic resources occupied increase due to the additional QDR memory controller and the notification service. However, the overhead is acceptable compared to the significant benefits of increased state storage.

When adding the statistics option in configuration c), the required logic resources increase again, but this time by a noticeable 50% compared to configuration b). In practice, tracking statistics is expensive due to the dedicated registers within each Handler and buffer module. Especially the implementation of the management access needs a large number of LUTs. The maximum frequency for all three configurations is determined by the longest path, which passes the highly parallelized IP checksum calculation function.

Comparing these results to related prior work is difficult due to the different features of comparable implementations (see also Section 4.4.6) and because public results of these implementations are scarce. However, DINI [Gro12a] reports a resource usage of 3,889 FFs and 6,885 LUTs for their single connection core with software-support for connection establishment and ARP / ICMP. Intilop [Int12b] gives only a number of 'less than 30,000' slices for their full featured hardware TCP core. NetStage in configuration a), which is the best comparable one, consumes 17,914 slices on the Xilinx Virtex 5 architecture for the fully mapped design. Finally, in relation to the feature set, the resource usage of NetStage is competitive with current commercial designs.

7.2.2 Handler Synthesis

To obtain application results, six different Handlers for the emulation of typical network services or actual vulnerabilities have been selected. All modules are working emulations of the described services, not simple examples, and have been used in a real datacenter environment (see Section 7.4). These six Handlers are

[Müh12b]:

- Web server: Imitates a webmail service running on a vulnerable web server (identified by a corresponding version header), collecting information about web server attacks.

- Telnet: Emulates a faux system administration CLI accepting any login / password to gather data about what combinations attackers try and the commands being executed after login. Together with the SMTP emulation the Telnet service contains the largest number of different dialog states.

- Mail server: Pretends to be an open relay simply accepting every mail (SMTP protocol) to gather information about spam attempts.

- MSSQL Slammer detection: Responds to MSSQL Ping and detects a malicious packet as sent by the Slammer [Moo03] worm.

- SMB login detection: Emulates the first steps of the protocol until client login. It is used to gather information about attack attempts on the SMB service.

- DNS server: Emulates a DNS server that resolves a single domain to collect information about DNS attacks.

All the Handlers have been automatically compiled using Malacoda. Table 7.2 lists the corresponding synthesis results.

Table 7.2: Synthesis results for compiled Handler modules

Handler	Opt.	LUT	Reg. Bits	BRAM	Max. Freq.
SMB	LUT	3,201	1,474	0	192 MHz
DNS	LUT	2,705	1,284	0	148 MHz
MSSQL (Slammer)	LUT	1,767	1,144	0	220 MHz
Telnet	LUT	3,817	1,663	0	189 MHz
Mail	LUT	2,304	1,413	0	198 MHz
Web	BRAM	2,302	1,210	4	225 MHz
Mail	BRAM	2,268	1,438	4	201 MHz
Web	LUT	5,553	1,204	0	196 MHz

On the LX240T device of the NetFPGA card, each Handler roughly requires around 2 to 4 percent of the device resources. As all Handlers are based on the same template (see Section 6.4), they are actually very close in size.

The number of Flip Flops differs only slightly between Handlers, as the compiler does not add additional pipeline registers. Variations in Flip Flops are due to flag registers for conditions and whether state data is used or not (in the latter case this part of the template code is optimized away by the compiler).

Variations in LUTs / BRAMs are due to the different number of implemented response packets and the complexity of the conditions. E.g., the MSSQL Slammer detection requires only few LUTs, as it does only contain a single response packet and a single condition (it only detects that particular attack). In contrast, the Telnet Handler has 570B of replies in 18 templates, together with corresponding conditions, so it requires more LUTs.

To evaluate the effect of LUT / BRAM optimization (see Section 6.4.6), the compiler has been extended to support forcing a particular implementation. This choice has been used to compile the Web and Mail Handler again with forcing the non-optimized variant, so that both can be compared. In the results, one can see the high area impact of template data when stored using LUTs. For the around 6600B of response templates contained within the Web server Handler, using BRAM is definitely the better option. Actually, for smaller storage sizes such as the around 250B for the Mail Handler, the choice of memory implementation does not have a major impact as the additional logic to control the BRAM has a similar area than the logic required to store the template bytes directly in LUTs.

7.2.2.1 Evaluation of Malacoda compiler

In contrast to the previously developed manual Handlers (see [Müh12c]), the compiled Handlers are larger. This is because of a change in the basic Handler template to allow for automatic compilation (e.g., the previous Handlers had a different state machine implementation which was not that generic), so that these Handlers cannot be compared directly. To evaluate the performance of the compiled Handlers in comparison with hand-crafted ones, a custom version of the current Web server Handler has been created that uses the compiler template but has been later manually optimized, e.g., conditions have been aggregated to reduce states.

The compiled version of the Web Handler is slightly larger compared to a manual implementation, but both achieve nearly the same performance when creating response packets (the number of cycles given in this example corresponds

Table 7.3: Comparison of compiled and manual Web Server Handler (BRAM version)

Handler	LUT	Reg. Bits	Max. Freq.	Cycles
Web-Basic (compiled)	1,861	1,293	224 MHz	84
Web-Basic (handmade)	1,254	1,136	226 MHz	77

to a 100B request generating a 1000B response). The compiled version requires additional cycles because of a more generic matching implementation. However, this could be improved with future versions of the compiler (e.g., by skipping condition evaluation when there is no further chance for a match).

The overhead in LUTs is due to a more complex regular expression matching implementation and additional logic in the generic implementation of output packet generation and template code for e.g., log packets. Since the compiler has been optimized for generating high-performance hardware for a broad spectrum of use-cases, the increased use of resources is acceptable here. The Handlers are still relatively small compared to the overall FPGA capacity and reliable allow reaching the performance goal.

Table 7.4: Lines of code for compiled Handlers

Handler	Malacoda LOC	VHDL LOC
SMB	63	1203
DNS	26	793
Telnet	79	2089
Web	55	1798
MSSQL (Slammer)	16	744
Mail	80	2453

As an indication for productivity, Table 7.4 lists the lines of codes that are required for the six Malacoda descriptions and the resulting VHDL code. This comparison shows a significant reduction in complexity. In practice, a new Handler could be developed in Malacoda in roughly half an hour, if the specification is known. And even if the generated code needs to be manually optimized later (which is still possible, as the generated VHDL is human readable), there is a huge saving due to the automatically generated template.

7.2.2.2 Impact of Handler data path width

Beyond operating frequency and parallelism of operations, the data path width is another major criterion for the achievable system throughput. However, a double width data path requires double the number of register bits in every processing stage. This also increases the number of routing resources to connect the different stages (and complicates component placement while meeting timing constraints), so that the overall resource requirements are highly dependent on this decision.

NetStage uses a 128b data path by default. To evaluate the impact of data path width for the Handlers, a variant of the current Web server Handler (BRAM variant) using just a 64b data path has been developed manually by modifying the compiled template (analogously to the previously done comparison in [Müh10b]). The results are given in Table 7.5. In practice, conversion between different data path widths in the core and the Handlers while maintaining the clock could be easily performed by the Slot wrappers.

Table 7.5: Synthesis results for 128b and 64b Web Server Handler (BRAM version)

Handler	LUT	Reg. Bits	Max. Freq.
Web-Basic (128 Bit)	1,861	1,293	224 MHz
Web-Basic (64 Bit)	1,732	1,048	224 MHz

Surprisingly, the area overhead of the 128b variant over the narrower 64b implementation is only of minor relevance here. This is different to previous results from [Müh10b]. The reason for this is, that in [Müh10b] the Handler was using LUTs to implement packet templates, giving the data path width a much larger impact on logic area than the BRAM storage used here. In the BRAM Handler, most of the logic is required by code that controls the Handler operation (read packet from buffer, write header, write control messages), so that the effect of reducing the data path width does not have that large an impact.

Based on these results, data path width reduction can be an option for LUT-based Handlers when Handler resources are scarce. However, LUT-based Handlers do only make sense for small handlers. In general, optimizations of the core template itself (see Table 7.3 above) promise larger area savings. Furthermore, a 64b data path comes at the cost of a 50% reduced interface throughput, which might need to be balanced by other optimizations if 10 Gbit/s are a hard requirement.

7.2.3 System results

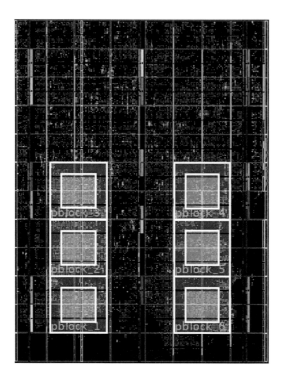

Figure 7.1: FPGA Layout for 6 VEH Slots

For the final mapped design, two different implementation options have been evaluated: monolithic and partitioned. The monolithic version contains all Handler modules in a single design without constraining their placement to particular FPGA partitions. The partitioned version introduces FPGA partitions for the Slots, so that Handlers can be replaced using partial reconfiguration.

For the partitioned design, the Slots have been floorplanned across the FPGA using PlanAhead as shown in Figure 7.1. The distribution has been chosen manually by inspecting previously implemented designs and locating suitable regions for the fixed-function network and memory interfaces, which require flexible placement for an efficient implementation. Each Slot partition contains 6400 LUTs/FFs and 8 BRAMs. The size of the Slots has been chosen to match the typical Handler sizes. In contrast to previous work [Müh11a], the Slots have now been sized somewhat larger here according to the slightly increased Handler size of the Malacoda compiled Handlers.

For this evaluation, configuration b) from Section 7.2.1 has been chosen (with SRAM enabled for GASM and notification service, but without statistics). Table 7.6 contains the final results for the two designs. As expected, the partitioned version requires slightly more resources. This is due to the more limited optimization that can be performed by the place & route process, as the Handler locations are constrained.

Table 7.6: Place & Route results for entire design including infrastructure (10G interface etc.) and six Handlers

Design	LUT	Reg. Bits	BRAM	Slices
monolithic	44,323	43,524	181	19,938
partitioned	47,881	43,845	181	20,690

In addition to the area results, another interesting aspect is the number of Slots that can be placed on an FPGA. The latest experiments with increasing Slot numbers on the NetFPGA card show as limiting factor not only the increasing requirement for LUTs, but also growing routing difficulties. The 10G interface blocks are very timing sensitive and require a particular placement, so that the two network interfaces reduce the available routing flexibility for the Slots. In practice, the NetFPGA platform can contain around six of the 6400 LUT/FF Slots before timing closure becomes too difficult.

7.2.4 Dynamic partial reconfiguration

The NetStage DPR setup is again based on configuration b) (see Section 7.2.1). As the core routing layer is already suitable for DPR, only the routing table and the management module have to be extended to allow self-adaptation by autonomous updates (see also Section 5.3). Bitfile storage is implemented using one of the QDR-SRAMs. Due to the limited amount of memory, the base design uses four Slots instead of six. Each Slot uses 512KB, so that bitfiles for four Handler/Slot combinations (16 bitfiles) will fit into the memory. RLDRAM would have been a better choice, but at the time of running the experiment the NetFPGA 10G project had not yet started work on the RLDRAM subsystem. But the SRAM-based storage is sufficient to evaluate the DPR technology. Please note that for practical experiments the BEE3 platform is better suited here, as the external SDRAM provides much more capacity. The DPR implementation works identically on both platforms.

Table 7.7 shows the results when implementing the design with Xilinx ISE 13.3. The results indicate that the DPR capability requires only a few hundred LUTs and register bits more in the management interface (for adaptation processes) than the non-DPR variant, and is a viable option for particular scenarios. The only constraint is that there needs to be available a suitable external memory for holding the bitstreams.

Table 7.7: Synthesis results for DPR components

Module	LUT	Reg. Bits	BRAM	Max. Freq.
NetStage Core	16,412	20,499	71	164 MHz
– Slot w/o Content	536	685	4	286 MHz
Mgmt., Routing, ICAP	4,432	5,037	61	201 MHz
NetStage (4 Slots)	**22,764**	**28,078**	**132**	**164 MHz**

7.2.4.1 Runtime Results

Of particular interest for the DPR scenario is the duration of a reconfiguration for a single Slot. At the standard ICAP speed of 100 MHz, it takes around $725\mu s$ to reconfigure a Slot with 6400 LUTs (283 KB). The precise reconfiguration time not only depends on the LUT size, but also on the placement of the partition, as Xilinx devices generally reconfigure based on frames (see [Fie10]), so that the given duration is valid only for a specific Slot layout (see Figure 7.1).

The reconfiguration time is stated excluding the time required to cleanly shutdown the slot, as this is dependent on the number of bytes that are still in the buffer and the processing speed of the Handler. The major limitation factor is the bandwidth provided by the 32b ICAP interface. As previously discussed, Duhem et al. [Duh11] have shown that it is actually possible to overclock the ICAP interface to a multiple of the vendor given 100MHz, but as this is beyond the device specification, it was not used here to prevent possible damage to the chip. The remaining control logic (e.g., prepare DMA transfer from QDR-SRAM to ICAP) does take a negligible amount of time. At the measured duration of $725\mu s$, a Handler could be replaced around 1380 times per second.

For applications such as the MalCoBox, this will be sufficient to adapt the system to incoming network traffic, as not every incoming packet will lead to a reconfiguration trigger event. Note that the process to schedule the next reconfiguration operation within the management module (see Section 5.3.3) requires only about 3 μs at 156.25 MHz clock speed (for a routing and counter table with 512 entries), so that this time is negligible.

At the other extreme, assuming an application where each arriving packet would generate a reconfiguration event, the overall achievable throughput would be reduced to around 5.5 MBit/s for 500B network packets. This shows that DPR is already well suited for particular scenarios, but true realtime reconfiguration still requires technological improvements (see also Section 2.1.4).

7.3 Network Performance

Results regarding the performance of the platform can be classified into the core performance and the specific handler performance. While the core performance (latency and throughput) is mostly independent from the application (except only the impact of GASM data forwarding if enabled for a particular handler), the performance of a single Handler depends highly on the implemented functionality.

7.3.1 Core Latency and Throughput

The core is designed to provide a throughput of at least 10 Gbit/s for network traffic that is assumed to consist of a mix of small and larger data packets. The throughput can be calculated by counting the cycles that are required to process a packet of a certain size. The total number of cycles consists of two different types: A fixed number of clock cycles overhead per packet for administrative functions (e.g., processing the internal control header data), and a variable number of clock cycles dependent on the payload size of the packet for content-related activities.

Figure 7.2 draws the throughput in relation to the packet size for the four major operations TCP Data Packet Receive, TCP Data Packet Send, UDP Data Packet Receive and UDP Data Packet Send. The calculations include preamble and inter-frame gap overhead as described in [Spi12] for a reference Ethernet speed of 10Gbit/s. Please further note that these performance numbers are guaranteed, as all processing blocks are implemented using dedicated (non-shared) resources that do not suffer from increased system load.

As expected, small packet sizes suffer from the administrative overhead cycles so that the overall throughput is adversely affected. Even if the modules use pipelining to a large extent, some administrative cycles cannot be avoided here because of NetStages's store-and-forward approach. E.g., the module has to wait for the result of a checksum calculation before it can forward the packet. But

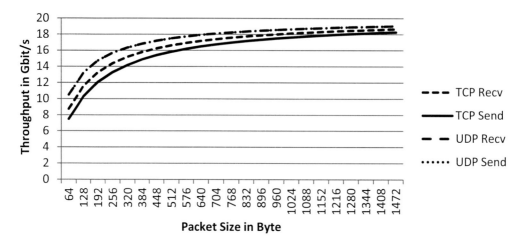

Figure 7.2: Network throughput for NetStage core operations

the value here is the greater flexibility, because the modules are decoupled and can be developed and exchanged independently. But overall, the graphs ramp up quickly, crossing the 10 Gbit/s target line and converge towards the internal data path maximum of 20 Gbit/s, so that the throughput goal is met.

Some commercial cores (see Section 4.4.6) perform better for small packets, as the data sheets indicate a throughput of 10 Gbit/s independent of the packet size. Again, this could be achieved by using a fully pipelined design through all stages (which will achieve better results for smaller packet sizes, as after reading a packet from the network the administrative overhead can be reduced by pipelining different operations).

In addition to throughput, the latency is another important characteristic of a network system. However, a comparison with existing work is difficult due to the lack of reliable published results. Only DINI reports information about the latency between the arrival of the first byte at the TCP offload core and the availability of the first byte on the application side (also named time to first byte, short TTFB) for a 100 byte packet payload with 120ns [Gro12b].

Because of the store-and-forward approach (see Section 4.2.2), the latency of a packet processed by NetStage applications depends on the size of the packet, and the number of stages that are traversed. In the best case scenario (all paths have empty buffers), the TTFB for a 100 byte TCP packet is 270ns, measured from the first byte entering the IP module until data is ready to leave the TCP module. While this is longer than the time on the DINI core, that core has been developed specifically for low latency and supports only a *single* connection. Furthermore, the DINI core processes IP and TCP in one step, because there is no need for decoupled IP processing, as performed by NetStage for greater modularity and flexibility. Despite this overhead, NetStage still manages to achieve a latency that is just 2.25x worse than DINI's.

To put this into perspective, note that on a regular Linux server system, even with a highly optimized network card and software stack such as Myricom DBL (Datagram Bypass Layer), the user-level read latency is about $1.5\mu s$ [Myr14]. This demonstrates the significant performance advantage of hardware-level network processing over purely software-based solutions.

7.3.2 Handler Performance

While the performance of the NetStage Core (latency and throughput) is mostly independent of the application, the performance of each single Handler does depend on its individual complexity. Table 7.8 shows the application-level throughput at the Handler in Gb/s, running at the current system target frequency of 156.25 MHz, and using the message sizes listed in brackets (excluding headers). The table lists the numbers once directly for the raw Handler operation and once in consideration of the in total 74 bytes minimal protocol headers that are added on the network layer (20 bytes TCP, 20 bytes IP, 14 bytes Ethernet and 20 bytes padding and inter-frame gap [Spi12]) for 10G Ethernet.

Table 7.8: Performance for example Handler operations

Handler Operation	Gb/s	
	direct	on wire
Web GET (78 bytes IN, 405 bytes OUT)	15.4	16.9
Send Mail (800 bytes IN, 14 bytes OUT)	16.9	18.1
Telnet uname (7 bytes IN, 19 bytes OUT)	2.9	9.6

As Table 7.8 shows, the constant number of overhead cycles becomes especially costly for very small payloads. However, only that specific Handler is slowed down, the NetStage core continues to process connections (including the TCP/IP stack) to other Handlers at the full speed of up to 20 Gb/s. Should tiny-payload performance be critical for certain applications, the limitation could be worked around by running multiple independent instances of the Handler, and performing load-balancing between them for higher aggregate throughput.

To stress-test the hardware in the lab environment, the Web Server Handler has been selected as robust testing tools are available for this type of service. The test has been performed on the BEE3 using the described test setup above (see Section 7.1.1). For 1 million of requests from 500 parallel clients sent with Apache Bench 2 [Apa10], the Handler replied in 22 μs (mean), while in comparison a software Apache required 100 μs (mean) to serve the same static page (fully loading all eight cores of the attached Linux test server) [Müh10c].

7.4 MalCoBox Live Test

While tests performed in a development environment are suitable to verify the basic functionality and to perform load testing, the interoperability with a broad range of different clients connecting from all over the world can only be simulated with great difficulty, using a huge range of test patterns. For this reason, in addition to the basic tests a live test has been set up with the MalCoBox system to verify the interoperability under real conditions. For this test, NetStage has been configured with the six compiled Handlers described above and connected to 10G Internet uplink at a German University data center using the NetFPGA 10G card plugged into a Linux server for management.

For this live test, configuration c) (see Section 7.2.1, QDR-SRAM to store application state data and the statistics option enabled) compiled as monolithic design has been used. The following repeats the results presented in [Müh12b], adding some additional information that has been gathered by analyzing the network dump of the original test run.

The original test has been run for one month. Two dedicated /25 subnets (= 252 usable IPs) have been assigned to the honeypot. The Handlers were configured to listen on all IP addresses. The MalCoBox was directly connected to the core router in the data center and had its own VLAN, so that the catchall-approach (see Section 4.3.2) could be easily applied. The public network traffic arriving at the public interface of the MalCoBox was dumped for later analysis on the management server using the NetStage mirroring option (see Section 5.2.3).

As a first result, this long-term evaluation proved the stability of the developed platform. No system failure or deadlock occurred during the entire test run.

7.4.1 Network Statistics

During the test period, 1.74 Million connection requests were reaching the Mal-CoBox honeypot. In that context, connection requests are defined as UDP packets or fully established TCP handshakes. This does not include packets arriving out of order, e.g., spoofed SYN-ACK responses. Table 7.9 lists the Top-10 services requested, as well as numbers for the remaining services for which the honeypot has active Handlers (active Handlers are shaded gray).

Table 7.9: Number of connections by service

Nr.	# Conn.	Port	Service
1.	977,549	445/TCP	MS-DS (SMB)
2.	167,430	80/TCP	HTTP
3.	82,882	139/TCP	NETBIOS Session
4.	36,167	3389/TCP	MS WBT Server
5.	31,093	1433/TCP	MS SQL Server
6.	30,966	8080/TCP	HTTP Alternate
7.	27,063	22/TCP	SSH
8.	20,118	23/TCP	Telnet
9.	15,618	210/TCP	Z39.50
10.	13,627	25/TCP	SMTP
44.	1838	1434/UDP	MS SQL Monitor
189.	243	53/UDP	DNS

©2012 ACM. Reprinted from [Müh12b].

Not surprisingly, with a connection rate of more than 50%, the Microsoft SMB protocol is leading the list. Due to its widespread use and various known vulnerabilities, SMB is a promising target for attackers (this is why Dionaea also primarily focuses on providing a SMB emulation (see Section 3.5.1)). In total, the four TCP-based Handlers are among the Top-10, such that the system had good coverage of network traffic.

The maximum number of newly established connections within a 300 seconds timeframe was 8000 (on average 25 req/s). The highest throughput measured was 25 MBit/s.

Actually, the number of 252 IP addresses might have been too low to achieve higher connection rates to really challenge the hardware (in a comparable software evaluation Baecher et al. [Bae06] used around 16,000). Increasing the number of IP addresses might be an option for future test runs, of course depending on availability.

7.4.1.1 TCP Traffic Characteristics

As one of NetStage's key components is a TCP implementation, the recorded traffic dump has been further analyzed to gather particular statistics about certain TCP stream characteristics. These are:

- TCP Retransmission: A packet has been retransmitted (from client to the MalCoBox) during an active connection.

- TCP Lost Segments: An expected segment of a connection has been finally lost.

- TCP Received Out of Order: TCP segments arrived out of order.

- Fragmented Flag Set: A packet has been fragmented into two or more fragments.

Analysis has been done using the tshark command line tool from wireshark [Wir11] in version 1.4.10. Table 7.10 lists the corresponding results.

Table 7.10: TCP traffic characteristics

Type	# Occurrences
Total Number of TCP Packets	6,585,885
TCP Retransmission	180,645 (2.74%)
TCP Lost Segments	44,610 (0.68%)
TCP Received Out of Order	8,516 (0.13%)
Fragmented Flag Set	0 (0%)

As one can see from the results, fragmentation indeed does not play any role for this type of communication (see also Section 4.3.3.2), which supports the decision of not investing in a implementation of fragmentation support. Furthermore, the number of packets received out of order is deceptively low, which leads to the conclusion that the measures implemented to avoid out of order transmissions did work well.

As expected, there is a small number of TCP retransmissions. Especially when accepting connections from all over the world, this is not avoidable, because packet loss of ACK messages cannot be prevented if packets travel multiple networks and carriers. There is also a small amounts of connections where even the retransmitted packets did never arrive, or no retransmission has been sent at all (e.g., because the client has terminated the sending thread or has been entirely shut down during the session), so that this will finally lead to a lost segment.

Finally, these results underline the statement that even the NetStage lightweight TCP implementation supports reliable communication on the Internet for research purposes.

7.4.1.2 Global Application State Memory Hash Access Key Distribution

Another result one can calculate from the traffic dump is the key distribution of the hash function used for access to the GASM content. To minimize the number of potential collisions, the key values should be equally distributed. For this analysis, the formula for the hash function has been implemented as C program and fed with the IP and port values extracted from the traffic dump for those packets, that would generate a GASM lookup (e.g., outgoing and TCP ACK messages have been ignored). Figure 7.3 shows the distribution of the 15 bit hash values that are used to address the GASM. In total, 866,694 unique IP / port combinations have been used to calculate the values.

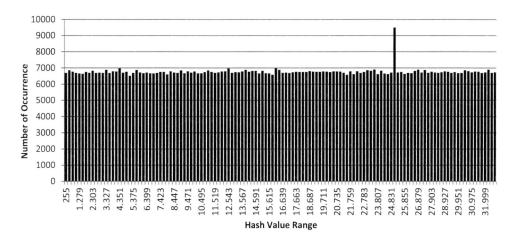

Figure 7.3: Distribution of hash values for GASM access

As one can see from the figure, the distribution is fairly equal amongst the range. Each bar in the figure contains the sum of hash values within a particular sub-range (the sub-range width is 256).

7.4.2 Service Emulations

In addition to counting the raw connections, monitoring milestones have been defined using the Malacoda **log** command to log when a certain step within the Handler has been reached [Müh12b]. The occurrence of these events is given in Table 7.11. Naturally these numbers are lower than the overall connection requests, as they represent specific dialog paths taken. Most of the remaining requests result from simple port scans that check for available services and version numbers, but do not actually initiate a more complex dialog.

Table 7.11: Statistics of monitoring events

Event	# Occurrences
Web server: GET URL	118,384
MSSQL: Slammer Worm	1,588
SMB server: Login Attempt	24,566
Mail server: Mail Queued	3,778
Telnet server: Login	11,438

©2012 ACM. Reprinted from [Müh12b].

7.4.2.1 Web server emulation

Around 70% of the clients that connected to the web server emulation were requesting a particular URL. The majority of these were attempts to reach a vulnerable service (e.g., phpMyAdmin) by sending a special crafted URL (obviously coming from automated attack scripts). Interestingly, clients were not actually trying to log into the webmail facade served by the Handler. It seems that such interactions are not the focus of the automatic scripts.

7.4.2.2 Mail server emulation

25% of the clients connecting to the mail service actually followed the entire emulated SMTP dialog to send a mail. These mails appear to be initial identification messages from spam engines, checking whether a server that looks like an open relay actually does deliver the mails. To avoid any adverse impact on the owner institution that provided the honeypot IP address range, these messages were not actually delivered. But it could be of particular interest for a further test run to actually deliver these emails selectively to induce attackers to send real spam.

7.4.2.3 Telnet server emulation

Clients that tried to log into the Telnet emulation used the username / password combination root / admin or simply root with no password in the majority of cases. The commands that were executed after the client was logged in were, e.g., echo test or echo connectioncheck, most likely coming from automatic scripts looking for open Telnet servers. As the Handler did not cover these commands, the emulation stopped here. For a future revision of the Telnet service, it might be of interest to implement the echo command as well.

7.4.2.4 SMB server emulation

While the majority of requests for the SMB service were simple port scans, some of the clients tried an anonymous login without password. But as further dialog paths have not been implemented in the Handler description, the emulation stopped here. Emulation of the SMB protocol is one of the more difficult ones, due to many variable length fields and partial encryption, which is not handled very efficiently by the current Malacoda compiler prototype. This is a point that could be addressed with future revisions of the compiler system.

7.4.2.5 Slammer worm detection

Even nine years after the large outbreak and the massive effort to remove the worm from infected systems, there has been a constant load of 20-40 Slammer requests per day trying to infect the honeypot. In a distributed and open worldwide network, such as the Internet, a threat like this is never disappearing once it has come up.

7.4.2.6 DNS server emulation

The numbers from Table 7.9 show only limited interest of attackers in the DNS server emulation. All requests appear to be coming from automatic vulnerability scanners and simply request arbitrary domain names. However, the traffic dump showed a huge number of responses arriving with spoofed IP addresses, obviously originating from DDoS attacks to real origin servers. This has been also observed for the HTTP service, where the dump showed massive numbers of TCP SYN ACK responses, that should not have been there as NetStage does not initiate connections.

7.5 Chapter Summary

This chapter presented evaluation results of an implementation of the described architecture and application. The NetStage components, the Malacoda compiler and the MalCoBox honeypot application have been benchmarked and verified in various experiments, using the BEE3 board and the NetFPGA 10G card as hardware base.

The synthesis results show that the footprint of the entire platform requires roughly one fifth of an Virtex 5 TX240 FPGA, the device family that is used on the NetFPGA 10G card. In comparison with commercial network communication cores, NetStage shows competitive results. Furthermore, the sizes of the compiled application Handlers for the honeypot are just 2 to 4 percent of such an FPGA, so that multiple Handlers fit on one FPGA. Furthermore, the simplified programming interface offered by Malacoda dramatically reduces development efforts for building new service emulations for the honeypot. The compiled handlers show similar performance than the hand-made ones.

In addition to initial tests performed in the lab environment, which demonstrated interoperability and stability, a live test of the hardware honeypot MalCoBox showed the usability of the system for this application under real-world conditions. Such a hardware honeypot can be operated just as any low-interaction software honeypot, collecting data unattended for long time periods, but without the risk of the system becoming compromised. In addition, the NetStage communication core exhibited its compatibility with many different client implementations from all over the world.

Summing up, the presented results clearly show the practical feasibility of the described architecture. It demonstrates a high potential in research, education and production environments.

8 Conclusions and Future Work

Security of computer systems and networks is one of the key issues for the future development of the Internet. But novel attack patterns and huge traffic rates overload purely software-based network security solutions. Therefore, hardware acceleration is employed for a number of purposes including deep packet inspection, encryption and firewalling. While such lower-layer implementations have been extensively studied in research and industry in the past, this work focuses on hardware for applications targeting *higher* communication layers. In this manner, this thesis opens up a promising direction for further research in that domain, targeting future high-speed networks, cloud applications and datacenter operation.

Such hardware-based applications benefit from two properties of dedicated architectures. First, custom hardware allows achieving superior performance for particular applications in contrast to a standard server. Second, if such a hardware-based application is implemented omitting a CPU, the vulnerability surface of the system is significantly reduced as there is no longer a general purpose CPU that can be subverted to run malicious code.

To explore the use of such hardware-based network applications in the security domain, a novel platform has been designed and developed as part of this thesis that allows rapid prototyping of FPGA-based applications that support autonomous Internet communication without any CPU involvement (see Chapter 4). This platform, called NetStage, contains a core implementation for the basic Internet communication protocols, a set of utility services, a flexible interface for application modules and tools to support development on the platform. In addition to this platform, this thesis also covered dynamic partial reconfiguration, a recent technology in FPGA design, to dynamically adapt the hardware structures to changing network conditions (see Chapter 5). The entire platform has been evaluated on the real-world example of an entire hardware-based honeypot application (see Section 7.4). NetStage demonstrates the high potential of hardware-accelerated networking even for complex scenarios such as the operation of a honeypot.

As a second topic beyond the hardware, this thesis also covers the aspect of *programming complexity*. FPGA designs traditionally have been programmed in hardware-related programming languages that follow different concepts than software-based systems, which make it difficult for developers familiar with general high-level programming languages, such as C or Java, to directly start working with FPGA designs. Using the example of the hardware honeypot, this thesis proposes the use of a domain-specific language, called Malacoda, to allow researchers and engineers to write application modules for the honeypot in an abstract way (see Chapter 6). This allows the complete hiding of the complexity of the underlying hardware. Together with the provided custom Malacoda compiler that generates high-performance hardware modules for NetStage, this flow forms an efficient way of working with the hardware-based honeypot system, for both hardware-experienced designers as well as domain experts without a background in digital systems design.

As a final proof-of-concept, the entire platform has been stress-tested during a live-test of the MalCoBox hardware honeypot that has been directly connected to a 10G Internet uplink at a German University datacenter (see Section 7.4). This test showed the practical functionality of the system interacting with different clients coming from all over the world. NetStage and the MalCoBox have also been used in student projects to develop e.g., a MySQL vulnerability emulation module for the MalCoBox. This also demonstrates the universal usability of the platform for research activity in that domain.

8.1 Summary and Conclusions

8.1.1 Platform

The hardware architecture of NetStage has been designed to be modular and flexible for extensions. The base component is the NetStage communication core (see Section 4.3). It supports fully autonomous Internet communication using the common TCP and UDP transport-layer protocols. To support different communication sessions and services on a single FPGA, the core contains a socket routing table that allows binding combinations of IP addresses and ports to hardware service modules. The current implementation follows a lightweight design (see Section 4.4) that aims at reducing the amount of session state information to a minimum to support millions of simultaneous connections at an aggregated speed of 10 Gbit/s.

For the implementation of application functionality the concept of Handlers has been introduced (see Section 5.1). Handlers are independent hardware service modules that perform a particular task. Multiple Handlers can exist in a system connected to the communication core using a shared bus infrastructure that has been combined with a message based communication layer to adapt to the networking characteristics of the data being transported. Handlers can use the services of the communication core to implement UDP / TCP endpoint functionality. In case of the honeypot application, the Handlers implement the (vulnerable) service emulations.

Beyond the core functionality, the platform also offers supporting services that aid in designing applications. These supporting services consist of a Global Application State Memory (GASM, see Section 4.2.6) and a Notification Timer (see Section 4.2.7). The GASM provides a limited amount of state memory (e.g., for Session-IDs or user credentials) at low latency for the itself memory-less Handlers. Data from this state memory is attached to every incoming packet, so that it is transparently available for processing by the corresponding Handler. The Notification Timer supports the injection of packets after a certain delay.

To complete the development platform, NetStage also offers extensive simulation capabilities (see Section 5.2.4). In addition to the standard test benches, there is also a real-time simulation interface that provides a virtual network interface, so that the simulation can be directly connected to network traffic. This allows quickly verifying and debugging the operation of new Handlers before running the entire hardware generation process.

8.1.1.1 Dynamic Partial Reconfiguration

To scale the available resources of a single chip beyond the physical limits, time-multiplexing of hardware-resources has been employed for the NetStage platform (see Section 5.3). NetStage implements DPR at the granularity of a single Handler. This allows long-term runs of experiments, where the FPGA hardware can adjust itself to changing traffic characteristics. E.g., in case of the honeypot, many Handlers can be held in an offline repository and instantly activated on request.

The process to reconfigure the Handler has been also implemented using dedicated hardware. DMA transfer is used to quickly feed the data from an external memory into the reconfiguration interface of the FPGA. Scheduling of reconfiguration activity has been implemented as a least-recently-used algorithm that counts packets arriving for any configured Handler in the routing table of the core. Periodically, Handlers with low activity are replaced by Handlers that show a higher activity count.

8.1.2 Malacoda

To address the common problem of missing high-level programming tools for reconfigurable technology, the domain-specific language Malacoda has been introduced to support the abstract description of Handler functionality for the honeypot domain (see Section 6.3). Malacoda allows describing protocol interactions, also called dialogs, on the network level.

Malacoda is influenced by the style service descriptions are commonly developed for software honeypots. It is based on Perl, a language that is very widespread in the networking and system administration domain, allowing domain experts to quickly write Malacoda descriptions. The entire hardware layer is hidden by Malacoda. There is no need to manually keep track of session state lookups or connection setup.

A compiler has been developed that converts Malacoda descriptions into hardware NetStage Handlers as VHDL modules (see Section 6.4). The compiler aims to achieve the same performance as the NetStage core by chaining commands to allow single-cycle processing of a 128b data word. This streaming-oriented principle is common in many hardware implementations and fully exploits the parallel processing capabilities of dedicated hardware, which would not be possible to implement using standard CPUs.

The compilation is performed by parsing the Malacoda description for commands and conditionals and grouping them into blocks that are then compiled into hardware blocks using a library of code templates (see Section 6.4). These hardware blocks are then wrapped into a Handler template that has been developed as a base skeleton and contains the interface logic for NetStage and the handling of the NetStage message format. The generation of VHDL as an intermediate representation has the further advantage, that experienced developers can easily modify the compiled Handler and add custom functionality, while they benefit from the quick generation of the base version of the Handler.

8.1.3 Results

The proposed concepts and the hardware architecture have been evaluated under real conditions using a NetFPGA 10G FPGA development board as base system. The already mentioned hardware honeypot, the MalCoBox, has been used as the evaluation prototype for any of the components. Using both handmade code and the Malacoda automatic compiler, six honeypot service Handlers have been developed, that emulate vulnerable services (Web, Mail, DNS, SMB, Telnet) or look for particular attacks (Slammer worm).

The synthesis results for the hardware (see Section 7.2) show that the core platform infrastructure requires roughly 30 percent of the Xilinx Virtex 5 FPGA provided on the NetFPGA 10G. As planned, this leaves sufficient free space for a set of Handlers to be integrated together with the core onto a single FPGA. The size of a single Handler ranges between 2 and 4 percent of the FPGA (see Section 7.2.2), while the compiled Handlers are only a bit larger than the hand-made ones. Furthermore, the performance of compiled Handlers does exceed the required 10 Gbit/s for the majority of Handlers. In consideration of the compiler being only a prototype, this is a very satisfying result.

The performance tests (see Section 7.3) showed that performance was sustained even under high load. This is as expected, as one advantage of dedicated hardware is that the dedicated processing elements assure a proper handling of requests even under load. While passing all the tests in the lab environment, the MalCoBox based on NetStage has also been evaluated in a real-world scenario.

The hardware honeypot has been connected to a 10 Gbit/s Internet uplink in a university datacenter and run for one month, collecting attack information from all over the world (see Section 7.4). In addition to collecting information about attacks being performed against the honeypot, this test also verifies the interoperability of the communication core with many different clients, having different network stack implementations and different types of connectivity. During that test, the system showed a solid behavior.

Finally, the entire platform and the MalCoBox application demonstrate the practical feasibility of such high-level network applications converted entirely into hardware appliances.

8.2 Future Work

Based on the experiences gained during platform development, the design of Malacoda, and the live evaluation, multiple directions for future research are possible that can be basically split into a hardware (NetStage) and a compiler (Malacoda) part.

8.2.1 Hardware

While the current version of the base platform contains everything to support experiments with autonomous Internet communication in today's networks on dedicated hardware, there are features that might be of interest for future work on that topic.

IPv6   With the upcoming relevance of IPv6 [Dee98], implementing support for this protocol will be surely an important improvement for future revisions of the communication core. Technically, this will basically include the extension of the IP address fields throughout the entire platform components to the 64 bit used by IPv6. Furthermore the Neighbor Discovery Protocol (NDP) needs to be implemented as the replacement for ARP.

SSL/TLS   Transport layer security (TLS) [Die08], formerly known as Secure Socket Layer, is a protocol to support encrypted and authenticated communication between endpoints on the Internet. It is the globally accepted standard for secure application communication and is used to primarily secure web sites, supported by many web browsers and web servers. As more and more applications will rely on encrypted communication, offering such a module as part of NetStage would be very interesting for future experiments.

   TLS consists of two major parts: session handshake and authentication as well as data encryption. For both parts, several security schemes are employed, that will have a major impact on the hardware implementation. Authentication makes use of Diffie-Hellmann and / or RSA for authenticated key exchange, while for the data encryption AES is the current cipher standard. An implementation will therefore require all of these algorithms as well as some control logic (see also [Iso10]).

Client Functionality   In the current version, the platform is targeting the server side of applications. While this has been decided here for particular reasons (see Section 4.2), it could be interesting for other use-cases to implement a client version as well, that initiates connections to remote stations. This could be, e.g., used to securely load malware that has been placed on web servers.

   Actually, when implementing such a client functionality, there should be some precautions to avoid overloading the target server systems, as the hardware might be powerful enough to bring down entire application clusters by just sending requests at line rate.

8.2.2 Compiler

The initial prototype of the Malacoda language and the compiler provided good results for the honeypot use case, thus validating the concept of high-level design. This offers a variety of interesting directions for future research.

Fundamentally, it would be interesting whether a more generic language can be established for describing such network-oriented applications that would then compiled to hardware. Activities on that field, such as PacketC (see Section 3.4.2), look promising and will be worth watching further.

Based on the existing Malacoda compiler, the following topics could be starting points towards a more general solution.

More Service Emulations   Currently there are six service emulations for the Mal-CoBox. As there are many more services and known vulnerabilities out there, it might be of interest for further live experiments to develop more Handlers. Ne-penthes [Bae06] or Dionaea [Dio11] might be good resources to start implementing other vulnerability emulations or server modules.

Extended Regular Expression Support   The current compiler supports a basic set of regular expressions required to model a variety of service emulations for the honeypot (see Section 6.3.3). However, regular expression support for hardware is a major research topic and there is ongoing work providing fast accelerators using dedicated hardware. It would make sense to do another review of existing solutions at some point in the future to check whether an existing solution can be integrated into the compiler.

Computations   Support for arithmetic computations will be important for services that generate some dynamic data (e.g., Session IDs or updated time stamps) from existing data. Such calculations should be supported at the same speed as basic operations. Furthermore, data values might require some further typification. This will also affect the handling of variables.

Access to External Memory   Currently, access to memory inside Handlers is very limited to a small amount of application state data sufficient for holding connection-based values. It could be required for future services to allow access to larger amount of (external) memory. Furthermore, it could be interesting to provide global memory for Handlers that is not bound to a single connection to store, e.g., statistical data or dynamic templates for responses.

Interface for Custom Modules   While the high-level compiler can span a wide range of applications, there will always be some cases where manual coding will be required. To ease this process, there should be a defined interface between the generated code and the manually inserted code. This should also include the possibility to seamlessly integrate the custom operations into the data flow of the entire module.

Template Resource Sharing   In the current compiler implementation, response packet templates are always stored once for each occurence in the Malacode description. However, an option could be to implement some type of resource sharing, so that the same text can be shared between different response packets within a single Handler. As response packets can consume large amounts of LUTs or BRAM content, this could be valuable for larger Handlers.

Optimized Condition Evaluation Phase   Normally, the entire input packet is read during condition evaluation to find matching conditions. However, if there are no conditions left that could be matched in later parts of the packet, reading the entire packet does waste cycles. Therefore, the condition evaluation section could be optimized so that the condition evaluation phase stops after the last word of *relevant* input data has been read. Note that the Handler could also skip initial data words if the data to be matched is located in later data words.

Bibliography

[Ach08]   Achronix, *Introduction to Achronix FPGAs*; URL `http://www.achronix.com`, last accessed 03.08.2012 (2008)

[Ach12a] Achronix, *SDK 1000-100G Development Card*; URL `http://www.achronix.com`, last accessed 27.03.2013 (2012)

[Ach12b] Achronix, *Speedster22i HD FPGA Family Datasheet, Rev. 1.6*; URL `http://www.achronix.com`, last accessed 03.08.2012 (2012)

[Ahm10]   Ali Ahmadinia, Josef Angermeier, Sandor P. Fekete, Tom Kamphans, Dirk Koch, Mateusz Majer, Nils Schweer, Jürgen Teich, Christopher Tessars and Jan C. Veen, *ReCoNodes Optimization Methods for Module Scheduling and Placement on Reconfigurable Hardware Devices*; In Dynamically Reconfigurable Systems, 199–221 (Springer Netherlands) (2010)

[Ala10]   Nikolaos Alachiotis, Simon A. Berger and Alexandros Stamatakis, *Efficient PC-FPGA Communication over Gigabit Ethernet*; In Proceedings of the 10th IEEE International Conference on Computer and Information Technology, 1727–1734 (IEEE Computer Society, Washington, DC, USA) (2010)

[Alb06]   Carsten Albrecht, Roman Koch and Erik Maehle, *DynaCORE: A Dynamically Reconfigurable Coprocessor Architecture for Network Processors*; In Proceedings of the 14th Euromicro International Conference on Parallel, Distributed, and Network-Based Processing, 101–108 (2006)

[Alb10]   Carsten Albrecht, Jürgen Foag, Roman Koch, Erik Maehle and Thilo Pionteck, *DynaCORE - Dynamically Reconfigurable Coprocessor for Network Processors*; In Dynamically Reconfigurable Systems, 335–354 (Springer Netherlands) (2010)

[Als09]   Faeiz Alserhani, Monis Akhlaq, Irfan U. Awan, John Mellor, Andrea J. Cullen and Pravin Mirchandani, *Evaluating Intrusion Detection Systems in High Speed Networks*; In Proceedings of the 5th. International Conference on Information Assurance and Security - Vol.02, 454–459 (2009)

[Apa10] Apache Foundation, *ab - Apache HTTP server benchmarking tool, v. 2.2.17*; URL `http://httpd.apache.org`, last accessed 12.04.2013 (2010)

[Atk10] Robert D. Atkinson, Stephen J. Ezell, Scott M. Andes, Daniel D. Castro and Richard Bennett, *The Internet Economy: 25 years after .com*; (The Information Technology & Innovation Foundation) (2010)

[Att06] M. Attig and G. Brebner, *Systematic Characterization of Programmable Packet Processing Pipelines*; In Field-Programmable Custom Computing Machines, 2006. FCCM '06. 14th Annual IEEE Symposium on, 195 –204 (2006)

[Att11] M. Attig and G. Brebner, *400 Gb/s Programmable Packet Parsing on a Single FPGA*; In Proceedings of the seventh ACM/IEEE Symposium on Architectures for Networking and Communications Systems, 12 –23 (2011)

[Atz10] Luigi Atzori, Antonio Iera and Giacomo Morabito, *The Internet of Things: A Survey*; Computer Networks, vol. 54(15):2787–2805 (Elsevier North-Holland, Inc., New York, NY, USA), URL `http://dx.doi.org/10.1016/j.comnet.2010.05.010` (2010)

[Bae06] Paul Baecher, Markus Koetter, Maximillian Dornseif and Felix Freiling, *The nepenthes platform: An efficient approach to collect malware*; In Proceedings of the 9th International Symposium on Recent Advances in Intrusion Detection, 165–184 (Springer) (2006)

[Bak06] Zachary Baker, *String matching architectures for network security on reconfigurable computers*; Ph.D. thesis, University of Southern California Los Angeles (2006)

[Bar08] Karel Bartoš, Martin Grill, Vojtech Krmícek, Martin Rehák and Pavel Celeda, *Flow Based Network Intrusion Detection System using Hardware-Accelerated NetFlow Probes*; In CESNET Conference 2008 : security, middleware, and virtualization - glue of future networks, 49–56 (CESNET, z. s. p. o, Prague) (2008)

[Bau09] Lars Bauer, Muhammad Shafique and Jörg Henkel, *MinDeg: a performance-guided replacement policy for run-time reconfigurable accelerators*; In Proceedings of the 7th IEEE/ACM international conference on Hardware/software codesign and system synthesis, 335–342 (ACM, New York, NY, USA) (2009)

[Bau12] L. Bauer, A. Grudnitsky, M. Shafique and J. Henkel, *PATS: A Performance Aware Task Scheduler for Runtime Reconfigurable Processors*; In Proceedings of the 20th Annual International Symposium on Field-Programmable Custom Computing Machines, 208 –215 (2012)

[Bec12] Christian Beckhoff, Dirk Koch and Jim Torresen, *Go Ahead: A Partial Reconfiguration Framework*; Proceedings of the Annual IEEE Symposium on Field-Programmable Custom Computing Machines, 37–44 (IEEE Computer Society, Los Alamitos, CA, USA) (2012)

[Bee09] *BEE3*; URL `http://www.becube.com`, ©BEEcube Inc. (2009)

[Bis06] J. Bispo, I. Sourdis, J.M.P. Cardoso and S. Vassiliadis, *Regular Expression Matching for Reconfigurable Packet Inspection*; In Proceedings of the IEEE International Conference on Field Programmable Technology (2006)

[Blo12] Michaela Blott, *FPGAs Head for the Cloud*; Xcell journal, vol. 80:20–23 (Xilinx) (2012)

[Blu10] Bluespec, *Bluespec vs. C/C++/SystemC Modeling - White Paper*; URL `http://www.bluespec.com`, last accessed 27.03.2013 (2010)

[Bra89] R. Braden, *Requirements for Internet Hosts - Communication Layers*; RFC 1122 (Standard) (IETF), URL `http://www.ietf.org/rfc/rfc1122.txt`, updated by RFCs 1349, 4379, 5884, 6093, 6298, 6633 (1989)

[Bra02] F. Braun, J. Lockwood and M. Waldvogel, *Protocol wrappers for layered network packet processing in reconfigurable hardware*; IEEE Micro, vol. 22(1):66 –74 (2002)

[Bre09] Gordon Brebner, *Packets everywhere: The great opportunity for field programmable technology*; Proceedings of the International Conference on Field Programmable Technology, 1–10 (2009)

[Bro10] Broadcom, *High Capacity StrataXGS® Trident II Ethernet Switch Series*; URL `http://www.broadcom.com`, last accessed 27.03.2013 (2010)

[Bro12] Broadcom, *NLA91024XT and NLA9512XT Proprietary Interface Knowledge-based Processors*; URL `http://www.broadcom.com`, last accessed 27.03.2013 (2012)

[Cal12]   Calypto Design Systems, Inc, *Catapult Product Family Datasheet*; URL `http://calypto.com`, last accessed 10.05.2014 (2012)

[CER12]   ICS CERT, *ICS CERT Monitor Q4 2012*; (2012)

[Cha10]   D. Chasaki and T. Wolf, *Design of a secure packet processor*; In Proceedings of the ACM/IEEE Symposium on Architectures for Networking and Communications Systems, 1 –10 (2010)

[Che06a] Sriram R. Chelluri, *Virtex-II Pro FPGAs Enable 10 Gb iSCSI/TCP Offload*; Storage & Servers Solution Guide (2006)

[Che06b] Sriram R. Chelluri, *Virtex-II Pro FPGAs Enable 10 Gb iSCSI/TCP Offload*; Xcell journal, vol. 57 (Xilinx) (2006)

[Che08]   Hao Chen and Yu Chen, *A Novel Embedded Accelerator for Online Detection of Shrew DDoS Attacks*; In Proceedings of the 2008 International Conference on Networking, Architecture, and Storage, 365–372 (2008)

[Che11]   Hao Chen, Yu Chen and D.H. Summerville, *A Survey on the Application of FPGAs for Network Infrastructure Security*; IEEE Communications Surveys Tutorials, vol. 13(4):541–561 (2011)

[Cis11a] Cisco, *Cisco Nexus 7000 Hardware Architecture*; BRKARC-3470 (2011)

[Cis11b] Cisco Systems, Inc., *NetFlow Version 9 Flow-Record Format*; URL `http://www.cisco.com/go/netflow`, [accessed 15 Jul 2011] (2011)

[Cla10]   Christopher Claus, Rehan Ahmed, Florian Altenried and Walter Stechele, *Towards Rapid Dynamic Partial Reconfiguration in Video-Based Driver Assistance Systems*; In Reconfigurable Computing: Architectures, Tools and Applications, vol. 5992 of *Lecture Notes in Computer Science*, 55–67 (Springer Berlin / Heidelberg) (2010)

[Clo08]   Cloudshield, *CS-2000 Content Processing Platform Data Sheet*; URL `http://www.cloudshield.com`, last accessed 18.01.2013 (2008)

[Cor09]   S. Corbetta, M. Morandi, M. Novati, M.D. Santambrogio, D. Sciuto and P. Spoletini, *Internal and External Bitstream Relocation for Partial Dynamic Reconfiguration*; IEEE Transactions on Very Large Scale Integration (VLSI) Systems, vol. 17(11):1650 –1654 (2009)

[Dav09]   John D. Davis, Charles P. Thacker and Chen Chang, *BEE3: Revitalizing Computer Architecture Research*; Tech. Rep. MSR-TR-2009-45, Microsoft Corporation (2009)

[Dav12]   Lucas Davi, Ra Dmitrienko, Manuel Egele, Thomas Fischer, Thorsten
          Holz, Ralf Hund, Stefan Nürnberger and Ahmad reza Sadeghi,
          *MoCFI: A framework to mitigate control-flow attacks on smart-
          phones*; In Proceedings of the Network and Distributed System
          Security Symposium (2012)

[DC09]   Lorenzo De Carli, Yi Pan, Amit Kumar, Cristian Estan and Karthikeyan
          Sankaralingam, *PLUG: Flexible Lookup Modules for Rapid Deploy-
          ment of New Protocols in High-speed Routers*; In Proceedings of
          the ACM SIGCOMM 2009 Conference on Data Communication,
          207–218 (ACM, New York, NY, USA) (2009)

[Dee98]   S. Deering and R. Hinden, *Internet Protocol, Version 6 (IPv6) Specifica-
          tion*; RFC 2460 (Draft Standard) (IETF), URL `http://www.ietf.`
          `org/rfc/rfc2460.txt`, updated by RFCs 5095, 5722, 5871, 6437,
          6564 (1998)

[Der10]   Luca Deri, Joseph Gasparakis, Peter Waskiewicz and Francesco Fusco,
          *Wire-speed hardware-assisted traffic filtering with mainstream net-
          work adapters*; (2010)

[Die08]   T. Dierks and E. Rescorla, *The Transport Layer Security (TLS) Protocol
          Version 1.2*; RFC 5246 (Proposed Standard) (IETF), URL `http:`
          `//www.ietf.org/rfc/rfc5246.txt`, updated by RFCs 5746, 5878,
          6176 (2008)

[Dio11]   Dionaea, *Dionaea Documentation*; URL `http://dionaea.carnivore.`
          `it`, last accessed 27.03.2013 (2011)

[Dol05]   Apostolos Dollas, Ioannis Ermis, Iosif Koidis, Ioannis Zisis and Christo-
          pher Kachris, *An Open TCP/IP Core for Reconfigurable Logic*;
          In Proceedings of the 13th Annual IEEE Symposium on Field-
          Programmable Custom Computing Machines, 297–298 (IEEE Com-
          puter Society) (2005)

[Duh11]   Francois Duhem, Fabrice Muller and Philippe Lorenzini, *FaRM: Fast
          Reconfiguration Manager for Reducing Reconfiguration Time Over-
          head on FPGA*; In Reconfigurable Computing: Architectures, Tools
          and Applications, vol. 6578 of *Lecture Notes in Computer Science*,
          253–260 (Springer Berlin / Heidelberg) (2011)

[Dun09]   R. Duncan and P. Jungck, *packetC Language for High Performance
          Packet Processing*; In Proceedings of the 11th IEEE International

Conference on High Performance Computing and Communications, 450 –457 (2009)

[Edd07] W. Eddy, *TCP SYN Flooding Attacks and Common Mitigations*; RFC 4987 (Informational) (IETF), URL `http://www.ietf.org/rfc/rfc4987.txt` (2007)

[Edw05] Stephen A. Edwards, *The Challenges of Hardware Synthesis from C-like Languages*; In Proceedings of the Design, Automation and Test in Europe Conference and Exhibition (DATE05) (2005)

[Fae09] Miad Faezipour, Mehrdad Nourani and Rina Panigrahy, *A hardware platform for efficient worm outbreak detection*; ACM Trans. Des. Autom. Electron. Syst., vol. 14(4):49:1–49:29 (ACM, New York, NY, USA) (2009)

[Fal00] Hamish Fallside and Michael J.S. Smith, *Internet Connected FPL*; In Field-Programmable Logic and Applications: The Roadmap to Reconfigurable Computing, vol. 1896 of *Lecture Notes in Computer Science*, 48–57 (Springer Berlin Heidelberg), URL `http://dx.doi.org/10.1007/3-540-44614-1_6` (2000)

[Fie10] Bruce Fienberg, *Xilinx Improves Design Flow for Industry's Only Proven Partial Reconfiguration FPGA Technology with ISE Design Suite 12.2*; URL `http://press.xilinx.com`, last accessed 27.03.2013 (2010)

[Gad07] H. Gadke and A. Koch, *Comrade - A Compiler for Adaptive Computing Systems using a Novel Fast Speculation Technique*; In Proceedings of the International Conference on Field Programmable Logic and Applications (2007)

[Gan10] Thilan Ganegedara, Yi-Hua E. Yang and Viktor K. Prasanna, *Automation Framework for Large-Scale Regular Expression Matching on FPGA*; In Proceedings of the International Conference on Field Programmable Logic and Applications, 50–55 (2010)

[Gar09] R. Garcia, A. Gordon-Ross and A.D. George, *Exploiting Partially Reconfigurable FPGAs for Situation-Based Reconfiguration in Wireless Sensor Networks*; In Field Programmable Custom Computing Machines, 2009. FCCM '09. 17th IEEE Symposium on, 243 –246 (2009)

[Get12] Jim Gettys and Kathleen Nichols, *Bufferbloat: Dark Buffers in the*

Internet; Commun. ACM, vol. 55(1):57–65 (ACM, New York, NY, USA) (2012)

[Gho11] Debasish Ghosh, *DSLs in Action*; 978-1935182450 (Manning) (2011)

[Gib08] G. Gibb, J.W. Lockwood, J. Naous, P. Hartke and N. McKeown, *NetF-PGA: An Open Platform for Teaching How to Build Gigabit-Rate Network Switches and Routers*; Education, IEEE Transactions on, vol. 51(3):364 –369 (2008)

[Gon05] I. Gonzalez, F. Gomez-Arribas and S. Lopez-Buedo, *Hardware-accelerated SSH on self-reconfigurable systems*; In Proceedings of the IEEE International Conference on Field-Programmable Technology, 289 – 290 (2005)

[Gop10] Lakshmi Gopalakrishnan, *XAPP853: QDR II SRAM Interface for Virtex-5 Devices*; Xilinx (2010)

[Gra10] Mentor Graphics, *Modelsim SE 6.6d Documentation*; (2010)

[Gre09] Peter Gregorius and Ulrich Langenbach, *High Speed Hardware Architectures: TCP/IP Stack*; URL `http://www.hhi.fraunhofer.de`, last accessed 27.03.2013 (2009)

[Gro12a] DINI Group, *TCP Offload Engine IP (TOE) Product Overview*; URL `http://www.dinigroup.com`, last accessed 27.03.2013 (2012)

[Gro12b] DINI Group, *TOE Latencies Product Brief*; URL `http://www.dinigroup.com`, last accessed 09.05.2013 (2012)

[Han11] S.G. Hansen, D. Koch and J. Torresen, *High Speed Partial Run-Time Reconfiguration Using Enhanced ICAP Hard Macro*; In Proceedings of the IEEE International Symposium on Parallel and Distributed Processing Workshops and Phd Forum, 174 –180 (2011)

[Hau99] Scott Hauck and William D. Wilson, *Runlength Compression Techniques for FPGA Configurations*; In Proceedings of the Seventh Annual IEEE Symposium on Field-Programmable Custom Computing Machines, 286– (IEEE Computer Society, Washington, DC, USA) (1999)

[Hau07] Scott Hauck and Andre DeHon, *Reconfigurable Computing: The Theory and Practice of FPGA-Based Computation (Systems on Silicon)*; (2007)

[Hay10]   David Hayes, Mattia Rossi and Grenville Armitage, *Improving DNS performance using Stateless TCP in FreeBSD 9*; Tech. rep., Centre for Advanced Internet Architectures (2010)

[He12]   Ke He, Louise Crockett and Robert Stewart, *Dynamic Reconfiguration Technologies Based on FPGA in Software Defined Radio System*; Journal of Signal Processing Systems, vol. 69:75–85 (Springer New York) (2012)

[Hop01]   J.E. Hopcroft, R. Motwani and J.D. Ullman, *Introduction to automata theory, languages, and computation*; Addison-Wesley series in computer science (Addison-Wesley) (2001)

[Hor08]   Yohei Hori, Akashi Satoh, Hirofumi Sakane and Kenji Toda, *Bitstream Encryption and Authentication Using AES-GCM in Dynamically Reconfigurable Systems*; In IWSEC '08: Proceedings of the 3rd International Workshop on Security, 261–278 (Springer-Verlag) (2008)

[Hur11]   Per Hurtig, Wolfgang John and Anna Brunstrom, *Recent Trends in TCP Packet-Level Characteristics*; In Proceedings of the Seventh International Conference on Networking and Services (2011)

[Hut02]   B.L. Hutchings, R. Franklin and D. Carver, *Assisting network intrusion detection with reconfigurable hardware*; In Proceedings of the 10th Annual IEEE Symposium on Field-Programmable Custom Computing Machines, 111 – 120 (2002)

[Int04]   Intel, *Intel IXP2800 Network Processor - Hardware Reference Manual*; (2004)

[Int12a] Intel, *Intel 82599 10 GbE Controller Datasheet, Rev. 2.76*; URL http://www.intel.com, last accessed 27.03.2013 (2012)

[Int12b] Intilop, *10 Gbit TCP Offload Engine (TOE) - Hardware IP Core - Top Level Product Specifications*; URL http://www.intilop.com, last accessed 17.08.2012 (2012)

[ISC12]   ISC, *Internet Host Count history*; URL https://www.isc.org, last accessed 22.08.2012 (2012)

[Iso10]   T. Isobe, S. Tsutsumi, K. Seto, K. Aoshima and K. Kariya, *10 Gbps implementation of TLS/SSL accelerator on FPGA*; In Proceedings of the 18th International Workshop on Quality of Service, 1 –6 (2010)

[IT10]   INVEA-TECH, *COMBO v2 Product Brief*; URL http://www.invea-tech.com, last accessed 26.03.2013 (2010)

[IT12] INVEA-TECH, *FlowMon Probe Product Brief*; URL `http://www.invea-tech.com`, last accessed 27.03.2013 (2012)

[Jan09] Hankook Jang, Sang-Hwa Chung and Dae-Hyun Yoo, *Design and implementation of a protocol offload engine for TCP/IP and remote direct memory access based on hardware/software coprocessing*; Microprocess. Microsyst., vol. 33(5-6):333–342 (Elsevier Science Publishers B. V., Amsterdam, The Netherlands, The Netherlands) (2009)

[Jed08] Gajanan S. Jedhe, *A Scalable High Throughput Firewall in FPGA*; In 16th International Symposium on Field-Programmable Custom Computing Machines (2008)

[Jia09] Weirong Jiang and Viktor K. Prasanna, *Large-scale wire-speed packet classification on FPGAs*; In Proceedings of the ACM/SIGDA international symposium on Field programmable gate arrays, 219–228 (ACM, New York, NY, USA) (2009)

[Jun12] Peder Jungck, Ralph Duncan and Dwight Mulcahy, *packetC Programming*; (2012)

[Kam02] Dan Kaminsky, *Paketto Keiretsu*; URL `http://www.doxpara.com`, last accessed 02.08.2012 (2002)

[Kar10] K. Karras, T. Wild and A. Herkersdorf, *A folded pipeline network processor architecture for 100 Gbit/s networks*; In Proceedings of the ACM/IEEE Symposium on Architectures for Networking and Communications Systems, 1 –11 (2010)

[Kat07] Toshihiro Katashita, Yoshinori Yamaguchi, Atusi Maeda and Kenji Toda, *FPGA-Based Intrusion Detection System for 10 Gigabit Ethernet*; IEICE Trans. Information and Systems, vol. E90-D:1923–1931 (2007)

[Kau11] Gaganpreet Kaur, *VHDL: Basics to Programming*; (Pearson Education) (2011)

[Kob09] P. Kobiersky, J. Korenek and L. Polcak, *Packet header analysis and field extraction for multigigabit networks*; In Proceedings of the 12th International Symposium on Design and Diagnostics of Electronic Circuits Systems, 96 –101 (2009)

[Koc08] Dirk Koch, Christian Beckhoff and Jürgen Teich, *ReCoBus-Builder a Novel Tool and Technique to Build Statically and Dynamically Reconfigurable Systems for FPGAs*; In Proceedings of International

Conference on Field-Programmable Logic and Applications, 119–224 (2008)

[Koc10] Andreas Koch, *Adaptive Computing Systems and Their Design Tools*; In Dynamically Reconfigurable Systems, 117–138 (Springer Netherlands) (2010)

[Koc11] Dirk Koch and Jim Torresen, *FPGASort: a high performance sorting architecture exploiting run-time reconfiguration on fpgas for large problem sorting*; In Proceedings of the 19th ACM/SIGDA international symposium on Field programmable gate arrays, 45–54 (ACM, New York, NY, USA) (2011)

[Kor10] Jan Korenek, *Fast Regular Expression Matching Using FPGA*; Information Sciences and Technologies Bulletin of the ACM Slovakia,, vol. 2:103–111 (2010)

[Kor11] Pavol Korcek, Vlastimil Kosar, Martin Zadnik, Karel Koranda and Petr Kastovsky, *Hacking NetCOPE to Run on NetFPGA-10G*; In Proceedings of the 2011 ACM/IEEE Seventh Symposium on Architectures for Networking and Communications Systems, 217–218 (IEEE Computer Society, Washington, DC, USA) (2011)

[Kos11] V. Kosar and J. Korenek, *Reduction of FPGA resources for regular expression matching by relation similarity*; In Proceedings of the IEEE 14th International Symposium on Design and Diagnostics of Electronic Circuits Systems, 401 –402 (2011)

[Krm11] Vojtech Krmicek, *Hardware-Accelerated Anomaly Detection in High-Speed Networks*; Ph.D. thesis, Masaryk University Faculty of Informatics (2011)

[Kul06] C. Kulkarni and G. Brebner, *Micro-Coded Datapaths: Populating the Space Between Finite State Machine and Processor*; In Proceedings of the International Conference on Field Programmable Logic and Applications, 1 –6 (2006)

[Lab09] Martin Labrecque, J. Gregory Steffan, Geoffrey Salmon, Monia Ghobadi and Yashar Ganjali, *NetThreads: Programming NetFPGA with Threaded Software*; In Proceedings of the NetFPGA Developers Workshop (2009)

[Lam07] J. Lamoureux and S.J.E. Wilton, *Clock-Aware Placement for FPGAs*; In

Proceedings of the International Conference on Field Programmable
Logic and Applications, 124 –131 (2007)

[Lav11] C. Lavin, M. Padilla, J. Lamprecht, P. Lundrigan, B. Nelson and
B. Hutchings, *HMFlow: Accelerating FPGA Compilation with Hard
Macros for Rapid Prototyping*; In Proceedings of the 19th Annual In-
ternational Symposium on Field-Programmable Custom Computing
Machines, 117 –124 (2011)

[Lav12] Maysam Lavasani, Larry Dennison and Derek Chiou, *Compiling high
throughput network processors*; In Proceedings of the ACM/SIGDA
international symposium on Field Programmable Gate Arrays, 87–96
(ACM, New York, NY, USA) (2012)

[Le10] Hoang Le and V.K. Prasanna, *High-throughput IP-lookup supporting
dynamic routing tables using FPGA*; In Proceedings of the International Conference on Field-Programmable Technology, 287 –290
(2010)

[Leb11] C. Leber, B. Geib and H. Litz, *High Frequency Trading Acceleration
Using FPGAs*; In Proceedings of the International Conference on
Field Programmable Logic and Applications, 317 –322 (2011)

[Lee04a] K. Lee, G. Coulson, G. Blair, A. Joolia and J. Ueyama, *Towards a
generic programming model for network processors*; In Proceedings
of the 12th IEEE International Conference on Networks, vol. 2, 504 –
510 vol.2 (2004)

[Lee04b] Sanghun Lee, Chanho Lee and Hyuk-Jae Lee, *A new multi-channel
on-chip-bus architecture for system-on-chips*; In Proceedings of the
IEEE International SOC Conference, 305 – 308 (2004)

[Lei05] Corrado Leita, Ken Mermoud and Marc Dacier, *ScriptGen: an automated script generation tool for honeyd*; In Proceedings of the 21st
Annual Computer Security Applications Conference, 203–214 (2005)

[Lit11] Heiner Litz, Christian Leber and Benjamin Geib, *DSL programmable
engine for high frequency trading acceleration*; In Proceedings of the
fourth workshop on High performance computational finance, 31–38
(ACM, New York, NY, USA) (2011)

[Liu09] M Liu, W. Kuehn, Z. Lu and A. Jantsch, *Run-time Partial Reconfiguration Speed Investigation and Architectural Design Space Ex-*

ploration; In Proceedings of the International Conference on Field Programmable Logic and Applications, 498–502 (2009)

[Lo09]   Chia-Tien Dan Lo and Yi-Gang Tai, *Space Optimization on Counters for FPGA-Based Perl Compatible Regular Expressions*; ACM Trans. Reconfigurable Technology and Systems, vol. 2(4):1–18 (2009)

[Loc07]   John W. Lockwood, Nick McKeown, Greg Watson, Glen Gibb, Paul Hartke, Jad Naous, Ramanan Raghuraman and Jianying Luo, *NetFPGA–An Open Platform for Gigabit-Rate Network Switching and Routing*; In Proc. of the 2007 IEEE International Conference on Microelectronic Systems Education, 160–161 (IEEE Computer Society) (2007)

[Loc11]   Kyle Locke, *XAPP1151: Content Addressable Memory Reference Design*; Xilinx (2011)

[Loc12]   John W. Lockwood, Adwait Gupte, Nishit, Michaela Blott, Tom English and Kees Vissers, *A Low-Latency Library in FPGA Hardware for High-Frequency Trading (HFT)*; In Proceedings of the 20th Annual Symposium on High-Performance Interconnects (2012)

[Loi07]   J. Loinig, J. Wolkerstorfer and A. Szekely, *Packet Filtering in Gigabit Networks Using FPGAs*; In Proceedings of the 15th Austrian Workshop on Microelectronics (2007)

[LSI10]   LSI, *Tarari T2000/T2500 Content Processor Product Brief*; (2010)

[Lyo09]   Gordon Lyon, *Nmap Network Scanning*; (Insecure Press) (2009)

[MA07]   M. Mirza-Aghatabar, S. Koohi, S. Hessabi and M. Pedram, *An Empirical Investigation of Mesh and Torus NoC Topologies Under Different Routing Algorithms and Traffic Models*; In Proceedings of the 10th Euromicro Conference on Digital System Design Architectures, Methods and Tools, 19 –26 (2007)

[Mah11]   P. Mahr, S. Christgau, C. Haubelt and C. Bobda, *Integrated Temporal Planning, Module Selection and Placement of Tasks for Dynamic Networks-on-Chip*; In Proceedings of the IEEE International Symposium on Parallel and Distributed Processing Workshops and Phd Forum, 258 –263 (2011)

[Mai09]   Gregor Maier, Anja Feldmann, Vern Paxson and Mark Allman, *On dominant characteristics of residential broadband internet traffic*;

In Proceedings of the 9th ACM SIGCOMM conference on Internet measurement, 90–102 (ACM) (2009)

[Mak04] Srihari Makineni and Ravi Iyer, *Architectural characterization of TCP/IP packet processing on the Pentium microprocessor*; In IEEE Proceedings on Software, 152 – 161 (2004)

[Mar08] T. Martmek and M. Kosek, *NetCOPE: Platform for Rapid Development of Network Applications*; In Proceedings of the 11th IEEE Workshop on Design and Diagnostics of Electronic Circuits and Systems, 1–6 (2008)

[Mar13] Marvell, *Marvell Xelerated HX300 Family of Network Processors*; URL `http://www.marvell.com`, last accessed 27.03.2013 (2013)

[Max09] Clive Maxfield, *FPGAs. World Class Designs*; (Newnes) (2009)

[Mee12] Wim Meeus, Kristof Van Beeck, Toon Goedemé, Jan Meel and Dirk Stroobandt, *An overview of today's high-level synthesis tools*; International Journal of Design Automation for Embedded Systems (DAES) (2012)

[Mer05] Marjan Mernik, Jan Heering and Anthony M. Sloane, *When and how to develop domain-specific languages*; ACM Comput. Surv., vol. 37(4):316–344 (ACM, New York, NY, USA), ©2005 ACM (2005)

[Mic09] Microsoft, *Microsoft SMB Protocol and CIFS Protocol Overview*; URL `http://msdn.microsoft.com`, last accessed 28.03.2013 (2009)

[Mog90] J.C. Mogul and S.E. Deering, *Path MTU discovery*; RFC 1191 (Draft Standard) (IETF), URL `http://www.ietf.org/rfc/rfc1191.txt` (1990)

[Moo03] D. Moore, V. Paxson, S. Savage, C. Shannon, S. Staniford and N. Weaver, *Inside the Slammer worm*; IEEE Security Privacy, vol. 1(4):33 – 39 (2003)

[Müh07] Sascha Mühlbach and Sebastian Wallncr, *Sccurc and Authenticated Communication in Chip-Level Microcomputer Bus Systems with Tree Parity Machines*; In Proceedings of the International Conference on Embedded Computer Systems: Architectures, Modeling and Simulation, 201 –208 (2007)

[Müh10a] Sascha Mühlbach, Martin Brunner, Christopher Roblee and Andreas Koch, *MalCoBox: Designing a 10 Gb/s Malware Collection Honeypot*

Using Reconfigurable Technology; Proceedings of the International Conference on Field Programmable Logic and Applications, 592–595 (IEEE Computer Society, Los Alamitos, CA, USA), ©2010 IEEE (2010)

[Müh10b] Sascha Mühlbach and Andreas Koch, *A Dynamically Reconfigured Network Platform for High-Speed Malware Collection*; In Proceedings of the International Conference on Reconfigurable Computing and FPGAs, 79 –84, ©2010 IEEE (2010)

[Müh10c] Sascha Mühlbach and Andreas Koch, *An FPGA-based scalable platform for high-speed malware collection in large IP networks*; In Proceedings of the International Conference on Field-Programmable Technology, 474 –478, ©2010 IEEE (2010)

[Müh11a] Sascha Mühlbach and Andreas Koch, *NetStage/DPR: A Self-adaptable FPGA Platform for Application-Level Network Security*; In Reconfigurable Computing: Architectures, Tools and Applications, vol. 6578 of *Lecture Notes in Computer Science*, 328–339 (Springer Berlin / Heidelberg), ©2011 Springer Berlin Heidelberg (2011)

[Müh11b] Sascha Mühlbach and Andreas Koch, *A novel network platform for secure and efficient malware collection based on reconfigurable hardware logic*; In Proceedings of the World Congress on Internet Security, 9 –14, ©2011 IEEE (2011)

[Müh11c] Sascha Mühlbach and Andreas Koch, *A Reconfigurable Hardware Platform for Secure and Efficient Malware Collection in Next-Generation High-Speed Networks*; International Journal for Information Security Research, vol. 1(4) (Infonomics Society), ©2011 Infonomics Society (2011)

[Müh11d] Sascha Mühlbach and Andreas Koch, *A Scalable Multi-FPGA Platform for Complex Networking Applications*; In Proceedings of the 19th Annual International Symposium on Field-Programmable Custom Computing Machines, 81 –84, ©2011 IEEE (2011)

[Müh12a] Sascha Mühlbach and Andreas Koch, *A Dynamically Reconfigured Multi-FPGA Network Platform for High-Speed Malware Collection*; International Journal of Reconfigurable Computing, vol. 2012 (2012)

[Müh12b] Sascha Mühlbach and Andreas Koch, *Malacoda: towards high-level compilation of network security applications on reconfigurable hard-*

ware; In Proceedings of the Symposium on Architecture for Networking and Communications Systems, ANCS '12, 247–258, ©2012 ACM (2012)

[Müh12c] Sascha Mühlbach and Andreas Koch, *NetStage/DPR: A self-reconfiguring platform for active and passive network security operations*; Microprocessors and Microsystems, ©2012 Elsevier B.V. (2012)

[Müh14] Sascha Mühlbach and Andreas Koch, *A Reconfigurable Platform and Programming Tools for High-Level Network Applications Demonstrated as a Hardware Honeypot*; IEEE Journal on Selected Areas in Communications, vol. 32(10):1919–1932, ©2014 IEEE (2014)

[Mye06] Glen Myers, *Overview of IP Fabrics PPL Language and Virtual Machine*; URL http://www.ipfabrics.com, last accessed 19.08.2012 (2006)

[Myk12] Mykonos, *Mykonos Web Security*; URL http://www.mykonossoftware.com, last accessed 02.08.2012 (2012)

[Myr12] Myricom, *Myricom 10G-PCIE-8B-S Adapter Datasheet*; URL https://www.myricom.com, last accessed 09.05.2013 (2012)

[Myr14] Myricom, *DBL Documentation and FAQ*; URL https://www.myricom.com, last accessed 10.05.2014 (2014)

[Nao08a] Jad Naous, David Erickson, G. Adam Covington, Guido Appenzeller and Nick McKeown, *Implementing an OpenFlow switch on the NetFPGA platform*; In Proceedings of the 4th ACM/IEEE Symposium on Architectures for Networking and Communications Systems, 1–9 (ACM, New York, NY, USA) (2008)

[Nao08b] Jad Naous, Glen Gibb, Sara Bolouki and Nick McKeown, *NetFPGA: reusable router architecture for experimental research*; In Proceedings of the ACM workshop on Programmable routers for extensible services of tomorrow, 1–7 (ACM, New York, NY, USA) (2008)

[Nee10] C. Neely, G. Brebner and Weijia Shang, *Flexible and Modular Support for Timing Functions in High Performance Networking Acceleration*; In Proceedings of the International Conference on Field Programmable Logic and Applications, 513 –518 (2010)

[Net11a] NetFPGA, *NetFPGA 10G loopback test example design*; URL https://github.com/NetFPGA, last accessed 27.03.2013 (2011)

[Net11b] NetFPGA, *NetFPGA 10G Public Wiki*; URL `https://github.com/NetFPGA`, last accessed 02.08.2012 (2011)

[Net12a] NetFPGA, *NetFPGA 10G Public Beta Announcement*; URL `http://netfpga.org`, last accessed 26.03.2013 (2012)

[Net12b] NetFPGA, *NetFPGA 1G/10G Project List*; URL `http://netfpga.org`, last accessed 26.03.2013 (2012)

[Net12c] F5 Networks, *SOL13876: Overview of the Auto Last Hop setting (11.x)*; URL `http://support.f5.com`, last accessed 18.06.2014 (2012)

[Neu08] Stephen Neuendorffer and Kees Vissers, *Streaming Systems in FPGAs*; In Embedded Computer Systems: Architectures, Modeling, and Simulation, vol. 5114 of *Lecture Notes in Computer Science*, 147–156 (Springer Berlin / Heidelberg) (2008)

[Ohl10] Rainer Ohlendorf, Michael Meitinger, Thomas Wild and Andreas Herkersdorf, *FlexPath NP Flexible, Dynamically Reconfigurable Processing Paths in Network Processors*; In Dynamically Reconfigurable Systems, 355–374 (Springer Netherlands) (2010)

[Onl12] Heise Online, *German EC cards: PINs can be stolen at card terminals*; URL `http://www.h-online.com`, last accessed 18.01.2013 (2012)

[Pan12a] Tian Pan, Xiaoyu Guo, Chenhui Zhang, Junchen Jiang, Hao Wu and Bin Liuy, *Tracking millions of flows in high speed networks for application identification*; In Proceedings of the IEEE INFOCOM, 1647 –1655 (2012)

[Pan12b] Pandalabs, *Pandalabs Security Report*; URL `http://www.pandalabs.com`, last accessed 26.03.2013 (2012)

[Par07] Terence Parr, *The Definitive Antlr Reference: Building Domain-Specific Languages*; URL `http://www.antlr.org/`, last accessed 02.08.2012 (2007)

[Pat07] Animesh Patcha and Jung-Min Park, *An overview of anomaly detection techniques: Existing solutions and latest technological trends*; Computer Networks, vol. 51(12):3448–3470 (Elsevier North-Holland, Inc., New York, NY, USA) (2007)

[Pej07] Vukasin Pejovic, Ivana Kovacevic, Slobodan Bojanic, Corado Leita, Jelena Popovic and Octavio Nieto-Taladriz, *Migrating a Honeypot to Hardware*; In Proceedings of the International Conference on

Emerging Security Information, Systems, and Technologies, 151–156 (IEEE Computer Society) (2007)

[Per12]   Perldoc, *Perl Language Reference v. 5.16.2*; URL `http://perldoc.perl.org`, last accessed 12.04.2013 (2012)

[Pey03]   Mohammad Peyravian and Jean Calvignac, *Fundamental architectural considerations for network processors*; (2003)

[Pio08]   Thilo Pionteck, Carsten Albrecht, Roman Koch and Erik Maehle, *Adaptive Communication Architectures for Runtime Reconfigurable System-on-Chips*; Parallel Processing Letters, vol. 18(02):275–289 (2008)

[PLD12]   PLDA, *QuickTCP Core - Reference Manual*; PLDA (2012)

[Plu82]   D. Plummer, *Ethernet Address Resolution Protocol: Or Converting Network Protocol Addresses to 48.bit Ethernet Address for Transmission on Ethernet Hardware*; RFC 826 (Standard) (IETF), URL `http://www.ietf.org/rfc/rfc826.txt`, updated by RFCs 5227, 5494 (1982)

[Pos80]   J. Postel, *User Datagram Protocol*; RFC 768 (Standard) (IETF), URL `http://www.ietf.org/rfc/rfc768.txt` (1980)

[Pos81a] J. Postel, *Internet Control Message Protocol*; RFC 792 (Standard) (IETF), URL `http://www.ietf.org/rfc/rfc792.txt`, updated by RFCs 950, 4884, 6633 (1981)

[Pos81b] J. Postel, *Internet Protocol*; RFC 791 (Standard) (IETF), URL `http://www.ietf.org/rfc/rfc791.txt`, updated by RFCs 1349, 2474 (1981)

[Pos81c] J. Postel, *Transmission Control Protocol*; RFC 793 (Standard) (IETF), URL `http://www.ietf.org/rfc/rfc793.txt`, updated by RFCs 1122, 3168, 6093, 6528 (1981)

[Pos83]   J. Postel, *The TCP Maximum Segment Size and Related Topics*; RFC 879 (IETF), URL `http://www.ietf.org/rfc/rfc879.txt`, updated by RFC 6691 (1983)

[Pro04a] Linux Virtual Server Project, *IPVS Kernel Module v. 1.2.1 Documentation*; URL `http://www.linuxvirtualserver.org`, last accessed 18.08.2012 (2004)

[Pro04b] Niels Provos, *A Virtual Honeypot Framework*; In In Proceedings of the 13th USENIX Security Symposium, 1–14 (2004)

[Pro07] Niels Provos and Thorsten Holz, *Virtual Honeypots: From Botnet Tracking to Intrusion Detection*; (Addison-Wesley Professional) (2007)

[Pro10] Fedora Project, *Fedora Core 14 Documentation*; URL `http://docs.fedoraproject.org`, last accessed 12.04.2013 (2010)

[Pta98] Thomas Ptacek, Timothy Newsham and Homer J. Simpson, *Insertion, Evasion, and Denial of Service: Eluding Network Intrusion Detection*; Tech. rep., Secure Networks, Inc. (1998)

[Qi10] Yaxuan Qi, J. Fong, Weirong Jiang, Bo Xu, Jun Li and V. Prasanna, *Multi-dimensional packet classification on FPGA: 100 Gbps and beyond*; In Proceedings of the International Conference on Field-Programmable Technology, 241 –248 (2010)

[Qi11] Yaxuan Qi, Kai Wang, J. Fong, Yibo Xue, Jun Li, Weirong Jiang and V. Prasanna, *FEACAN: Front-end acceleration for content-aware network processing*; In Proceedings of the 30th IEEE International Conference on Computer Communications, 2114 –2122 (2011)

[Roe99] Martin Roesch, *Snort - Lightweight Intrusion Detection for Networks*; In Proceedings of the 13th USENIX conference on System administration, 229–238 (1999)

[Ros04] Daniel L. Rosenband, *The ephemeral history register: flexible scheduling for rule-based designs*; In MEMOCODE, 189–198 (IEEE) (2004)

[Rot08] Nadav Rotem, *C to Verilog web service*; URL `http://www.c-to-verilog.com/`, last accessed 27.03.2013 (2008)

[Rub10] Erik Rubow, Rick McGeer, Jeff Mogul and Amin Vahdat, *Chimpp: a click-based programming and simulation environment for reconfigurable networking hardware*; In Proceedings of the 6th ACM/IEEE Symposium on Architectures for Networking and Communications Systems, 36:1–36:10 (2010)

[Sad11] Rabah Sadoun, *An FPGA based soft multiprocessor for DNS/DNSSEC authoritative server*; Microprocessors and Microsystems, vol. 35 (2011)

[Sai11] Paramjot Saini, Mandeep Singh and Balwinder Singh, *VHDL Implementation of PCI Bus Arbiter Using Arbitration Algorithms*; In

Contemporary Computing, vol. 168 of *Communications in Computer and Information Science*, 559–560 (Springer Berlin Heidelberg) (2011)

[San09] Oliver Sander, Benjamin Glas, Christoph Roth, Jürgen Becker and Klaus D. Müller-Glaser, *Priority-based packet communication on a bus-shaped structure for FPGA-systems*; In Proceedings of the Conference on Design, Automation and Test in Europe, 178–183 (European Design and Automation Association, 3001 Leuven, Belgium, Belgium) (2009)

[Sch02] D.V. Schuehler and J. Lockwood, *TCP-Splitter: A TCP/IP flow monitor in reconfigurable hardware*; In Proceedings of the 10th Symposium on High Performance Interconnects, 127 – 131 (2002)

[Sch04] David Vincent Schuehler, *Techniques for processing TCP/IP Flow Content in Network Switches at Gigabit Line Rates*; Ph.D. thesis, Sever Institute of Washington University (2004)

[sG09]   secXtreme GmbH, *honeyBox Honeypot Appliance*; URL `http://www.sec-xtreme.com`, last accessed 27.03.2013 (2009)

[Shi05] Alan Shieh, Andrew C. Myers and Emin Gün Sirer, *Trickles: a stateless network stack for improved scalability, resilience, and flexibility*; In NSDI'05: Proceedings of the 2nd conference on Symposium on Networked Systems Design & Implementation, 175–188 (USENIX Association) (2005)

[Sid01] R. Sidhu and V.K. Prasanna, *Fast Regular Expression Matching Using FPGAs*; In Proceedings of the 9th Annual IEEE Symposium on Field-Programmable Custom Computing Machines, 227 – 238 (2001)

[Sir08] Scott Sirowy, *Where's the Beef? Why FPGAs Are So Fast*; Tech. rep., Microsoft Research (2008)

[Soh11] A.A. Sohanghpurwala, P. Athanas, T. Frangieh and A. Wood, *OpenPR: An Open-Source Partial-Reconfiguration Toolkit for Xilinx FPGAs*; In Proceedings of the IEEE International Symposium on Parallel and Distributed Processing Workshops and Phd Forum, 228 –235 (2011)

[Son05] Haoyu Song and John W. Lockwood, *Efficient packet classification for network intrusion detection using FPGA*; In Proceedings of the 2005

ACM/SIGDA 13th international symposium on Field-programmable gate arrays, 238–245 (ACM, New York, NY, USA) (2005)

[Son11] Nehir Sonmez, Oriol Arcas, Gokhan Sayilar, Osman Unsal, Adrin Cristal, Ibrahim Hur, Satnam Singh and Mateo Valero, *From Plasma to Bee-Farm: Design Experience of an FPGA-Based Multicore Prototype*; In Reconfigurable Computing: Architectures, Tools and Applications, vol. 6578 of *Lecture Notes in Computer Science*, 350–362 (Springer Berlin / Heidelberg) (2011)

[Sop12] Sophos, *Security Threat Report 2013*; Last accessed 08.01.2013 (2012)

[Sou04] Ioannis Sourdis and Dionisios Pnevmatikatos, *Pre-Decoded CAMs for Efficient and High-Speed NIDS Pattern Matching*; In Proc. 12th Annual Symposium on Field-Programmable Custom Computing Machines, 258–267 (2004)

[Sov09] C. Soviani, I. Hadzic and S.A. Edwards, *Synthesis and Optimization of Pipelined Packet Processors*; IEEE Transactions on Computer-Aided Design of Integrated Circuits and Systems, vol. 28(2):231 –244 (2009)

[SPE02] SPECTER, *SPECTER Intrusion Detection System - Markers*; URL `www.specter.com`, last accessed 02.08.2012 (2002)

[Spi12] Spirent, *How to test 10 Gigabit Ethernet Performance (White Paper), Rev. B*; URL `http://www.spirent.com`, last accessed 18.06.2014 (2012)

[Sta04] Xilinx Staff, *Celebrating 20 Years of Innovation*; XCell48 (2004)

[Ste04] C. Steiger, H. Walder and M. Platzner, *Operating systems for reconfigurable embedded platforms: online scheduling of real-time tasks*; IEEE Transactions on Computers, vol. 53(11):1393 – 1407 (2004)

[Sut04] Alistair Sutcliffe and Nikolay Mehandjiev, *End-User Development*; Communications of the ACM, Vol. 47 No. 9 (2004)

[Sym12] Symantec, *Internet Threat Report 2011*; URL `http://www.symantec.com` (2012)

[Syn12] Synopsys, Inc, *Synphony C Compiler Datasheet*; URL `http://www.synopsys.com`, last accessed 10.05.2014 (2012)

[Tab10] Tabula, *Spacetime 3D Architecture*; URL `http://www.tabula.com`, last accessed 27.03.2013 (2010)

[Tab11]   Tabula, *NetASAP Brochure*; URL `http://www.tabula.com`, last accessed 27.03.2013 (2011)

[Tan10]   Z. Tan, K. Asanovi and D. Patterson, *An FPGA-based Simulator for Datacenter Networks*; In Proceedings of the Exascale Evaluation and Research Techniques Workshop at the 15th International Conference on Architectural Support for Programming Languages and Operating Systems (2010)

[tcp10]   tcpdump, *tcpdump v. 4.1.1 Documentation*; URL `http://www.tcpdump.org/`, last accessed 12.04.2013 (2010)

[Tig03]   Hans Tiggeler, *Modelsim SE UART-Socket FLI Demo*; URL `http://www.ht-lab.com`, last accessed 22.04.2013 (2003)

[Unn11]   D. Unnikrishnan, J. Lu, Lixin Gao and R. Tessier, *ReClick - A Modular Dataplane Design Framework for FPGA-Based Network Virtualization*; In Proceedings of the 7th ACM/IEEE Symposium on Architectures for Networking and Communications Systems, 145–155 (2011)

[Vai11]   Nilay Vaish, Thawan Kooburat, Lorenzo De Carli, Karthikeyan Sankaralingam and Cristian Estan, *Experiences in Co-designing a Packet Classification Algorithm and a Flexible Hardware Platform*; In Proceedings of the 2011 ACM/IEEE Seventh Symposium on Architectures for Networking and Communications Systems, 189–199 (IEEE Computer Society, Washington, DC, USA) (2011)

[Vig09]   Giovanni Vigna, Fredrik Valeur, Davide Balzarotti, William K. Robertson, Christopher Kruegel and Engin Kirda, *Reducing errors in the anomaly-based detection of web-based attacks through the combined analysis of web requests and SQL queries*; Journal of Computer Security, vol. 17(3):305–329 (2009)

[Vu11]   Tran Huy Vu, Nguyen Quoc Tuan, Tran Ngoc Thinh and Nguyen Tran Huu Nguyen, *Memory-efficient TCP reassembly using FPGA*; In Proceedings of the Second Symposium on Information and Communication Technology, 238–243 (ACM, New York, NY, USA) (2011)

[Wag05]   Arno Wagner and Bernhard Plattner, *Entropy Based Worm and Anomaly Detection in Fast IP Networks*; In Proceedings of the 14th International Workshops on Enabling Technologies: Infrastructure for Collaborative Enterprise, 172–177 (2005)

[Wag11] Cynthia Wagner, Jérôme François, Radu State and Thomas Engel, *Machine learning approach for IP-flow record anomaly detection*; In Proceedings of the 10th International IFIP TC 6 Conference on Networking, 28–39 (2011)

[Wan10] Hao Wang, Shi Pu, Gabriel Knezek and Jyh-Charn Liu, *A modular NFA architecture for regular expression matching*; Proceedings of the International Symposium on Field Programmable Gate Arrays, 209–218 (2010)

[Wan11] Kai Wang, Yaxuan Qi, Yibo Xue and Jun Li, *Reorganized and Compact DFA for Efficient Regular Expression Matching*; In Proceedings of the IEEE International Conference on Communications, 1 –5 (2011)

[Wea07] Nicholas Weaver, Vern Paxson and Jose M. Gonzalez, *The shunt: an FPGA-based accelerator for network intrusion prevention*; In Proceedings of the 2007 ACM/SIGDA 15th international symposium on Field programmable gate arrays, 199–206 (ACM, New York, NY, USA) (2007)

[Wic06] Georg Wicherski, *Medium Interaction Honeypots*; (2006)

[Wic10] Georg Wicherski, *Mwcollectd: Placing a low-interaction honeypot in-the-wild: A review of mwcollectd*; Network Security, vol. 2010(3):7–8 (Elsevier Science Publishers B. V., Amsterdam, The Netherlands, The Netherlands) (2010)

[Wil04] David Wile, *Lessons learned from real DSL experiments*; Sci. Comput. Program., vol. 51(3):265–290 (Elsevier North-Holland, Inc., Amsterdam, The Netherlands, The Netherlands) (2004)

[Wir11] Wireshark, *tshark Documentation v. 1.4.10*; URL `https://www.wireshark.org`, last accessed 13.05.2013 (2011)

[Wu94] Yu-Liang Wu and Douglas Chang, *On the NP-completeness of Regular 2-D FPGA Routing Architectures and a Novel Solution*; 362–366 (IEEE Computer Society Press, Los Alamitos, CA, USA), URL `http://dl.acm.org/citation.cfm?id=191326.191492` (1994)

[Xil10a] Xilinx, *UG086: Memory Interface Solutions*; (2010)

[Xil10b] Xilinx, *UG150: LogiCORE IP XAUI*; v9.2 edn. (2010)

[Xil11a] Xilinx, *UG081: MicroBlaze Processor Reference Guide*; (2011)

[Xil11b] Xilinx, *UG148: LogiCORE IP 10-Gigabit Ethernet MAC User Guide*; (2011)

[Xil11c] Xilinx, *UG626: Xilinx ISE 13.3 - Synthesis and Simulation Design Guide*; (2011)

[Xil11d] Xilinx, *UG695: ISE In-Depth Tutorial*; (2011)

[Xil11e] Xilinx, *UG702: Partial Reconfiguration User Guide*; (2011)

[Xil11f] Xilinx, *UG761 - AXI Reference Guide*; (2011)

[Xil12a] Xilinx, *UG175: LogiCORE IP FIFO Generator User Guide*; (2012)

[Xil12b] Xilinx, *UG190: Virtex-5 FPGA User Guide*; (2012)

[Xil12c] Xilinx, *UG191: Virtex-5 FPGA Configuration User Guide*; (2012)

[Xil12d] Xilinx, *UG902: Vivado High Level Synthesis User Guide*; (2012)

[Xil12e] Xilinx, *Xilinx Investor Factsheet - First Quarter Fiscal Year 2013*; (2012)

[Xil14]   Xilinx, *DS190: Zynq-7000 All Programmable SoC Overview*; (2014)

[Yam08] N. Yamagaki, R. Sidhu and S. Kamiya, *High-speed regular expression matching engine using multi-character NFA*; In Proceedings of the International Conference on Field Programmable Logic and Applications, 131 –136 (2008)

[Yan08] Yi-Hua E. Yang, Weirong Jiang and Viktor K. Prasanna, *Compact architecture for high-throughput regular expression matching on FPGA*; In Proceedings of the 4th ACM/IEEE Symposium on Architectures for Networking and Communications Systems, 30–39 (ACM, New York, NY, USA) (2008)

[Yan12] Yi-Hua Yang and V.K. Prasanna, *High-Performance and Compact Architecture for Regular Expression Matching on FPGA*; IEEE Transactions on Computers, vol. 61(7):1013 –1025 (2012)

[Yu04]   Fang Yu, *Gigabit Rate Packet Pattern-Matching Using TCAM*; In Proceedings of the 12th IEEE International Conference on Network Protocols (ICNP04) (2004)

[Yua10] Ruan Yuan, Yang Weibing, Chen Mingyu, Zhao Xiaofang and Fan Jianping, *Robust TCP Reassembly with a Hardware-Based Solution for Backbone Traffic*; In Proceedings of the IEEE Fifth International Conference on Networking, Architecture and Storage, 439 –447 (2010)

[Zei05]   A.S. Zeineddini and K. Gaj, *Secure partial reconfiguration of FPGAs*;
           In Proceedings of the IEEE International Conference on Field-
           Programmable Technology, 155 –162 (2005)

[Zet10]   Kim Zetter, *Report Details Hacks Targeting Google, Others*; URL
           `http://www.wired.com/threatlevel/2010/02/apt-hacks/`, last
           accessed 08.01.2013 (2010)

[Zie95]   G. Ziemba, D. Reed and P. Traina, *Security Considerations for IP
           Fragment Filtering*; RFC 1858 (Informational) (IETF), URL `http:
           //www.ietf.org/rfc/rfc1858.txt`, updated by RFC 3128 (1995)

List of Figures

List of Tables